INTEGRATING INNOVATION AND TECHNOLOGY MANAGEMENT

ETM WILEY SERIES IN ENGINEERING & TECHNOLOGY MANAGEMENT

Series Editor: Dundar F. Kocaoglu

PROJECT MANAGEMENT IN MANUFACTURING AND HIGH TECHNOLOGY OPERATIONS
Adedeji B. Badiru, University of Oklahoma

MANAGERIAL DECISIONS UNDER UNCERTAINTY: AN INTRODUCTION TO THE ANALYSIS OF DECISION MAKING
Bruce F. Baird, University of Utah

INTEGRATING INNOVATION AND TECHNOLOGY MANAGEMENT
Johnson A. Edosomwan, IBM Corporation

CASES IN ENGINEERING ECONOMY
Ted Eschenbach, University of Missouri-Rolla

MANAGEMENT OF RESEARCH AND DEVELOPMENT ORGANIZATIONS
Ravinder K. Jain, U.S. Army Corps of Engineers
Harry C. Triandis, University of Illinois

STATISTICAL QUALITY CONTROL FOR MANUFACTURING MANAGERS
William S. Messina, IBM Corporation

KNOWLEDGE BASED RISK MANAGEMENT IN ENGINEERING: A CASE STUDY IN HUMAN-COMPUTER COOPERATIVE SYSTEMS
Kiyoshi Niwa, Hitachi, Ltd.

MANAGING TECHNOLOGY IN THE DECENTRALIZED FIRM
Albert H. Rubenstein, Northwestern University

MANAGEMENT OF INNOVATION AND CHANGE
Yassin Sankar, Dalhousi University

PROFESSIONAL LIABILITY OF ARCHITECTS AND ENGINEERS
Harrison Streeter, University of Illinois at Urbana-Champaign

INTEGRATING INNOVATION AND TECHNOLOGY MANAGEMENT

JOHNSON A. EDOSOMWAN
IBM Corporation

A WILEY-INTERSCIENCE PUBLICATION
JOHN WILEY & SONS
NEW YORK   CHICHESTER   BRISBANE   TORONTO   SINGAPORE

PROTECTION OF VITAL TECHNICAL INFORMATION

No portion of this material may be reproduced in part or in full without written permission from the author and John Wiley & Sons. All written correspondence should be directed to the author at the address specified below:

Dr. Johnson A. Edosomwan
IBM Corporation
General Products Division
Department 737, Building 123
5600 Cottle Road
San Jose, CA 95193
(408)-256-7680

Copyright © 1989 by John Wiley & Sons, Inc.

All rights reserved. Published simultaneously in Canada.

Reproduction or translation of any part of this work beyond that permitted by Sections 107 or 108 of the 1976 United States Copyright Act without the permission of the copyright owner is unlawful. Requests for permission or further information should be addressed to the Permission Department, John Wiley & Sons, Inc.

Library of Congress Cataloging in Publication Data:
Edosomwan, Johnson Aimie,
   Integrating innovation and technology management / Johnson A. Edosomwan.
      p. cm.– (Wiley series in engineering & technology management)
   "A Wiley-Interscience publication."
   Bibliography: p.
   ISBN 0-471-61699-0
   1. Technological innovations–Management. 2. Technology–Management. I. Title. II. Series: Wiley series in engineering and technology management.
   HD45.E28   1989
   658. 5'14–dc19   89-5689
                                                                      CIP

Printed in the United States of America

10 9 8 7 6 5 4 3 2 1

To My Family

CONTENTS

Preface xv
Acknowledgments xvii

1 Basic Concepts in Innovation and Technology Management 1

   1.1 The Importance of Technological Innovation to Society, 1
   1.2 Definition of Invention and Innovation, 3
       1.2.1 Invention, 3
       1.2.2 Innovation, 3
   1.3 Innovation Process Model (IPM), 6
   1.4 Technology Trends, 10
   1.5 Impact of Technological Progress, 12
   1.6 Definition of Innovation and Technology Management, 15
   1.7 Management of Technology (MOT), 15
       1.7.1 Managing Technology Changes and Discontinuities, 16
   1.8 S-Curves, 18
   1.9 Technology Management Challenges, 20
       1.9.1 Managing Technology's Impact on Management, 20
       1.9.2 Managing Technology's Impact on Workers and Professionals, 21
       1.9.3 Managing the Shrinking Technology Life Cycle, 21
       1.9.4 Providing Communication Ability for Technologies Under One System Network, 22

1.9.5 Managing the Human Total Reliance on Technology, 22
1.9.6 Designing Acceptable Technologies, 23
1.9.7 Managing the Introduction of New Technologies into the Work Environment, 23
1.9.8 Managing the Impact of Technology on the Quality of Working Life, 23
1.9.9 Managing Technology Systems Requirements, 23
1.9.10 Managing Information, 23
1.9.11 Managing the Impact of Technology on Productivity and Quality, 24
1.9.12 Managing Workers in a High-Technology Environment, 24
1.10 Understanding Technology Outcome Dimensions, 24
Summary, 26
Questions and Problems, 26

2 Requirements for Innovation and Technology Management 29

2.1 Management's Role in Innovation and Technology Issues, 29
2.2 Formulating the Right Technological Policy, 30
2.3 Promoting Innovation through Cooperation Among Industry, Government, and Academia, 33
2.4 Promoting Innovation Through Technical Information Exchange, 34
2.5 Environmental Requirements for Encouraging Individual Innovativeness, 35
2.6 Individual Innovators, 36
2.7 The Ideal Innovative Manager, 36
2.8 Preparing for the Next-Generation Technology, 38
 2.8.1 Impact of Poor Technology Strategy, 39
 2.8.2 Recommended Approach to Preparing for Next-Generation Technology, 39
 2.8.3 Requirements for Managing Current- and Next-Generation Technology, 40
Summary 41
Questions and Problems, 41

3 Comprehensive Design Rules for Emerging Technology 43

3.1 Technology-Aided Systems Design Optimization Cycle, 43
 3.1.1 System Information-Gathering Phase, 43
 3.1.2 System Planning, Assessment, and Measurement Phase, 45
 3.1.3 System Selection Phase, 45
 3.1.4 System Testing and Improvement Phase, 45
3.2 Design Rules for Appropriate Technology, 46
 3.2.1 Design Rule 1: People are Center of Design, 46

3.2.2   Design Rule 2: Utilize Principle of Kinesiology in the Design, 46
3.2.3   Design Rule 3: Use Physiological Criteria Design, 51
3.2.4   Design Rule 4: Apply Psychological Principles to Improve Morale and Increase Job Satisfaction, 51
3.2.5   Design Rule 5: Recognize the Worker's Right over Control of Work Activity, 53
3.2.6   Design Rule 6: Allow for Skill Development through the Use of New Technology, 56
3.2.7   Design Rule 7: Design User-Friendly Systems, 56
3.2.8   Design Rule 8: Involve the Users of Technology-based Systems during the Design Phase, 56
3.2.9   Design Rule 9: Give the Technology-based Systems Ability to Advise, Alert, or Warn Users of Potential Events, 56
3.2.10 Design Rule 10: Implement A Total Productivity and Quality Improvement System, 57
3.3 Understanding Technology's Impact on the Quality of Working Life, 58

Summary, 59
Questions and Problems, 59

4 Justifying and Implementing Technology   61

4.1 Understanding the Uncertainties in Capital Decisions, 61
4.2 A Perspective on the Technology Justification Problem, 62
4.3 The Payback Period Justification Technique, 63
4.4 The Return on Investment (ROI) Justification Technique, 64
4.5 Net Present Value (NPV) Justification Technique, 65
4.6 Internal Rate of Return (IRR) Justification Technique, 66
4.7 Total Productivity Justification Technique, 67
4.8 Equal Productivity Curve Justification Technique, 71
4.9 Cost Curves Justification Technique, 71
4.10 Scoring Models for Justifying Technology, 72
     4.10.1 Unweighted Scoring Model, 72
     4.10.2 Weighted Scoring Model, 72
4.11 Allocating Resources to Different Technology Choices, 73
4.12 Ten-Step Approach for Implementing a New Technology, 75
4.13 The 6C Principles for Managing Technology Research and Development Projects, 76
     4.13.1 Principle One: Provide Controls, 77
     4.13.2 Principle Two: Provide a Focal Point Coordination, 77
     4.13.3 Principle Three: Provide Adequate Communication Channels, 77

4.13.4 Principle Four: Provide Adequate Focus on Cost Avoidance, 77
4.13.5 Principle Five: Implement Measures to Analyze the Contribution of Each Phase, 77
4.13.6 Principle Six: Facilitate Cooperation among Project Participants, 78
4.14 Common Implementation Problems, 78
4.15 Overall Strategy for Overcoming Common Problems, 81
4.16 Measuring Technology Progress, 85
Summary, 87
Questions and Problems, 88

5 Managing Technology Utilization, Transfer, and Forecasting, 91

5.1 Managing Technology Utilization, 91
5.2 Managing Technology Transfer, 93
  5.2.1 Obstacles to Technology Transfer, 93
  5.2.2 Methods of Technology Transfer, 94
5.3 Managing Technological Forecasting, 96
  5.3.1 Definition of Technological Forecasting, 96
  5.3.2 Operational Definition of Technological Forecasting by Levels in High-Technology Industry, 97
  5.3.3 Technological Forecasting Methodology in High-Technology Industry, 97
  5.3.4 Problems and Pitfalls in Technological Forecasting, 101
  5.3.5 Keys to the Successful Use of the Technological Forecast, 102
  5.3.6 Key Issues in Technological Forecasting, 102
Summary, 103
Questions and Problems, 103

6 Managing Product Design and Development Activities 105

6.1 Characteristics of Research And Development Organizations, 105
6.2 Matrix Organizational Model, 107
6.3 Flexed Contingency Leadership Style, 108
6.4 Designing a Product for Automation and Production, 109
  6.4.1 Benefits of Product Design for Automation, 109
  6.4.2 Product Life Cycle and Ideal Manufacturing Goals, 110
  6.4.3 Factors Affecting Product Design for Automation, 111
  6.4.4 Minimizing the Impact from Product Design for Automation on Manufacturing Processes, 111
  6.4.5 Application of Value Analysis Technique in Designing Product for Automation, 112
6.5 The Group Technology Concept, 114
6.6 Product Release Methods, 114

6.7 Early Manufacturing Involvement (EMI) in the Technology Design Phase, 117
    6.7.1 EMI Defined, 117
    6.7.2 EMI Objectives, 119
    6.7.3 Matrix Organizational Model for EMI, 120
    6.7.4 Leadership Style for EMI Projects, 120
    6.7.5 Communication in the EMI Organization, 120
Summary, 121
Questions and Problems, 121

7 Managing Current-Generation Technology — 123

7.1 Knowledge-Based Expert Systems, 123
    7.1.1 KBES Definition, Structure, and Configuration, 123
    7.1.2 KBES Impact on Productivity, 124
    7.1.3 Design Rules, 125
    7.1.4 Managing a KBES Project, 127
    7.1.5 KBES Applications Opportunities, 131
7.2 Robotics, 131
7.3 Computer-Aided Design (CAD) and Computer-Aided Manufacturing (CAM), 134
7.4 Computer-Integrated Manufacturing (CIM), 135
7.5 Information-Based Organization Structures, 135
    7.5.1 TINOP Layer One: Nonprogrammed Information and Decision Layer, 135
    7.5.2 TINOP Layer Two: Programmed Information and Decision Layer, 136
    7.5.3 TINOP Layer Three: Information and Decision Delivery Layer, 136
    7.5.4 Handling TINOP Challenges, 136
7.6 Areas of Rapidly Changing Technology, 138
Looking Ahead, 142
Questions and Problems, 142

8 Case Studies in Technology Management — 143

8.1 Case Study One: Problems of Technology Leadership—Xerox, Apple, and IBM—and Markets for Radical Innovation, 143
    8.1.1 Problems of Technology Leadership—Xerox, Apple, and IBM, 143
    8.1.2 Markets for Radical Innovation, 146
    8.1.3 IBM PC, 146
8.2 Case Study Two: Computer-Integrated Manufacturing Implementation at a General Electric Plant, 148
    8.2.1 Introduction, 148

- 8.2.2 A Lighting Panelboard, 148
- 8.2.3 Summary: A-Series Lighting Panelboard Project, 149
- 8.2.4 Product Design for CIM 150
- 8.2.5 Manufacturing Technology/Processes and CIM, 152
- 8.2.6 FFS Control System, 153
- 8.2.7 Plant Operations and Workforce, 153
- 8.2.8 Payoff, 154
- 8.3 Case Study Three: Choosing Among Competing Technologies—A Study of Automatic Identification Technologies, 155
  - 8.3.1 Introduction, 155
  - 8.3.2 Evaluation Methodology, 156
  - 8.3.3 Selection Among Automatic Identification Technology Alternatives, 158
  - 8.3.4 Summary and Discussion, 162
- 8.4 Case Study Four: Ergonomic Issues in Technology Application, 162
  - 8.4.1 Research Site and Methodology, 162
  - 8.4.2 Description of the Computer-Aided Task, 163
  - 8.4.3 The Subjects and Research Sequence, 164
  - 8.4.4 Measurement Methods for Variables, 165
  - 8.4.5 Results, 168
  - 8.4.6 Conclusions, 168
- 8.5 Case Study Five: Technology Impact on the Quality of Working Life, 169
  - 8.5.1 Situational Analysis of Technology Impact on the Quality of Working Life, 169
  - 8.5.2 Results and Discussion, 169
- 8.6 Case Study Six: Justifying the Use of Robotics, 170
  - 8.6.1 Research Site and Methodology, 170
  - 8.6.2 Description of the Manual Method of Assembling Printed Circuit Boards, 170
  - 8.6.3 Description of the Robotic Device Method of Assembling Printed Circuit Boards, 170
  - 8.6.4 Measurement Method for Productivity, 171
  - 8.6.5 Implementation Methodology, 171
  - 8.6.6 The Project Manager's Role during Implementation, 174
  - 8.6.7 Results, 174
  - 8.6.8 Conclusions, 176
- 8.7 Case Study Seven: Productivity Measurement in a Group Technology (GT) Production Environment, 177
  - 8.7.1 Total Productivity Measurement in a GT Environment, 177
  - 8.7.2 Machine and Part Grouping Techniques, 178
  - 8.7.3 Allocation Criteria for Overhead Expenses, 178

        8.7.4 Study Results, 179
        8.7.5 Problems Encountered during Implementation, 182
        8.7.6 Conclusions, 183
    8.8 Case Study Eight: Implementing Knowledge-Based Expert Systems, 183
        8.8.1 Operation Sequence, 184
        8.8.2 Case Study Results, 185
        8.8.3 Conclusions, 186
    8.9 Case Study Nine: A Productivity Perspective to Manage Technology Changes, 187
        8.9.1 Application of CTPM to Manage Technology Discontinuities, 189
        8.9.2 Example, 189
        8.9.3 Advantages of the CTPM Perspective, 191
    Acknowledgments, 192

9 Case Studies on Systems Integrators and Computer-Integrated Manufacturing Systems   193

    9.1 Introduction and Description of the Environment, 193
        9.1.1 Emergence of the Systems Integrator, 194
        9.1.2 Role of the MSI, 194
    9.2 Problem Focus and Methodology, 195
        9.2.1 Sample of Entrepreneurial MSIs, 195
        9.2.2 Data Collection Methods, 196
    9.3 Results, 197
        9.3.1 Successful Cases, 197
        9.3.2 Failure Cases, 199
    9.4 Case Analyses and Discussion of Results, 200
        9.4.1 Conditions Found to Promote Adoption, 201
        9.4.2 Conditions Found to Retard Adoption, 203
    9.5. Recommendations, 209
    9.6 Conclusions, 212
    9.7 Challenges for Further Work, 214
    Summary, 218
    Acknowledgments, 218
    Questions, 218

10 Technology Management Case Study: General Electric High-Speed Horizontal Project   219

    10.1 Background, 219
    10.2 The Winds of Change, 222

10.3 The Lighting Leadership Program, 224
10.4 The Manufacturing Technology Programs Department (MTPD), 226
10.5 The High-Speed Horizontal Project, 229
10.6 Running the HSH, 235
10.7 The Next Steps, 236
Acknowledgments, 239
Questions, 239

11 Roles and Responsibilities in New Product Development   **241**

11.1 Introduction, 241
11.2 Roles and Responsibilities, 242
11.3 Design Feasibility, 242
11.4 Product Feasibility, 247
11.5 Manufacturing Feasibility, 250
11.6 Design/Manufacturing/Marketing Interfaces, 251
11.7 Implementation, 253
Summary, 253
Acknowledgments, 253
Questions, 253

Conclusion   **255**

References   **261**

Further Reading   **269**

Appendix A   **Computer-Aided Manufacturing Ergonomic Checklist (CAMEC)**   **283**

Appendix B   **Comparative Judgment Instrument (CJI) for Assessment of Job Satisfaction and Psychological Stress**   **285**

Index   **291**

PREFACE

In the past four decades, technological innovation has been one of the major driving forces behind the improvement in productivity and the quality of working life. Technological advancement worldwide in tools, techniques, and procedures has helped the human race to perform simple, complex, and difficult operations that were once impossible. In a competitive world economy, the organizations that have instituted an effective system for encouraging and managing technological innovation are in a better position to produce better goods and services that will compete effectively in the world market.

This book was written with the belief that practical and integrative approaches are required for managing innovation and technology at the enterprise level. The book provides essential tools, techniques, and methodologies that are required to manage the interactive, integrated process of innovation and technology issues. The book presents new but tested ideas for handling technology management requirements and problems. It also assembles useful approaches for managing innovation and technology that are recommended by other experts and practicing managers in the field.

In Chapter 1, the terminology and concepts involved in the management of technological innovation are discussed. The innovation process model, technology management essentials, and the technology-aided task impact model are presented. Innovation and technology management challenges are presented, as are strategies for addressing them. Chapter 2 discusses in detail the requirements for managing technological innovation and technology issues at the firm level. A strategy for formulating an effective innovation and technology management program is presented. The characteristics of ideal innovative managers, innovators, and their environments are discussed.

Chapter 3 presents a comprehensive design rule for emerging technologies. The technology-aided design optimization cycle is also presented. Also discussed is strategy for fostering better relationships among the designers, developers, and users of technology. Chapter 4 presents several new techniques and methodologies for justifying and implementing a new technology. The technology-oriented total productivity model is presented. The intangible factors that need to be addressed during the justification process for a new technology are discussed, as are the implementation approach for a new technology in the workplace and the common problems experienced. The principles of control, coordination, cooperation, contribution analysis, communication, and cost avoidance for managing technology research and development projects are discussed. A framework for managing technology utilization, transfer, and forecasting is presented in Chapter 5. Chapter 6 provides guidelines for designing a product for automation and production. The value engineering technique for reducing the total number of parts required in a given product is presented. The product life cycle and the ideal manufacturing goals are discussed. The concepts of group technology, early manufacturing involvement are discussed along with application in a manufacturing process.

In Chapter 7 specific ideas on how to prepare for the next generation of technology are offered. A discussion of some of the most commonly used current technologies is presented. Chapters 8, 9, and 10 present technology management case studies from real-life experiences of major firms to illustrate the applicability of the various techniques presented. Chapter 11 presents the roles and responsibilities of new products development. The appendices present self-assessment checklists and instruments for technology management. An extensive bibliography on innovation and technology management is also offered.

This book evolved out of research, teaching, consulting, conducting seminars, conferences and workshops nationally and internationally, and several years of industrial work experience in the area of innovation and technology management.

The book is intended to serve the practical needs of practitioners, researchers, consultants, students, managers, executives, and government officials concerned with the process of managing and improving technological innovation. It is valuable as a reference book for class instruction or as a reference handbook for addressing innovation and technology management issues.

JOHNSON A. EDOSOMWAN

San Jose, California
August 1989

ACKNOWLEDGMENTS

Many organizations and people contributed to this book. My thanks to the Social Science Research Council (SSRC), U. S. Department of Labor, and International Business Machines Corporation (IBM) for their Grant Numbers SS-36-83-21 and IBM-2J2-722271-83/85. SSRC and IBM grants helped in conducting research work in technology management, texting the technology-aided task impact model (TATIM), and in gathering data for some of the case studies. My gratitude to all those companies and organizations that provided their facilities as a testing ground for methodologies prescribed in this book. For proprietary reasons the identities of some of the companies and organizations are masked. I am also grateful for the training and support received from my employer, IBM. My thanks to Jane Barnhart, Joelle Stufano, and Laura Degnan for the typing assistance they provided during the preparation of this book. My gratitude to Dr. William. E. Souder, W. J. Sheeran, T. R. Campbell, J. C. Sutton, S. McConnell, Dr. Paul Adler, and Dr. David J. Sumanth for their contributions to the various case studies presented in this book. My appreciation is expressed to my family who supported my early education. My love and appreciation to my wife, Mary, daughters, Esosa and Efe and son, Johnson Aimie Edosomwan, Jr. for their support and encouragement. The peaceful and innovative home life provided was very helpful. Finally, I am grateful to God for His guidance during the preparation of this book and for answering my prayers during moments of frustration.

INTEGRATING INNOVATION AND TECHNOLOGY MANAGEMENT

1
BASIC CONCEPTS IN INNOVATION AND TECHNOLOGY MANAGEMENT

This chapter presents the essential concepts and definitions in innovation and technology management. A framework and model for enhancing the innovation process are presented. Key process elements involved in innovation and technology management are discussed, as are mechanisms for stimulating the field of technology management. In the discussion of challenges and opportunities in innovation and technology management, the views of other experts presented in the literature are offered. A framework for understanding the overall impact of technology in the work environment is also provided.

1.1 THE IMPORTANCE OF TECHNOLOGICAL INNOVATION TO SOCIETY

Technological innovation is the process of creating and implementing new technology, products, and production and service capabilities. It allows society to get more from the same stock of resources because it provides a means for the conversion of society's stock of resources into goods and services, and for the sale or exchange of these goods and services in the market. The evidence shows that the productivity levels and the quality of goods and services produced by a firm can be significantly affected by the rate of technological change and the types of technologies used over time.

Abernathy and Clark (1985) stated that innovations may alter product design, production systems, skills and knowledge base, materials, and capital equipment. In the marketing area, innovations may alter customer bases, customer applications, channels of distribution and service, customer knowledge, and modes of communication. Porter (1985) emphasized that technological change is one of the principal drivers of competition. It plays a major role in structural change of existing industry, as well as in creation of new industries. It is also a great equalizer, eroding the competitive advantage of even well-entrenched firms and propelling others to the front. Frohman (1981) recommended that business firms should in fact use technology as a weapon to retain their competitive edge. Abernathy (1982) and Lewis (1982) have shown that improvement or decline in industrial productivity can be correlated with changes in the rate of technological change. Firms that have instituted an effective system for encouraging and managing technological innovation are in a better position to develop, manufacture, and provide products or services.

The products and services obtained through technological innovation may also provide excellence in the workplace and in society as a whole. Technological advances in tools, techniques, and procedures help extend the capacity of the

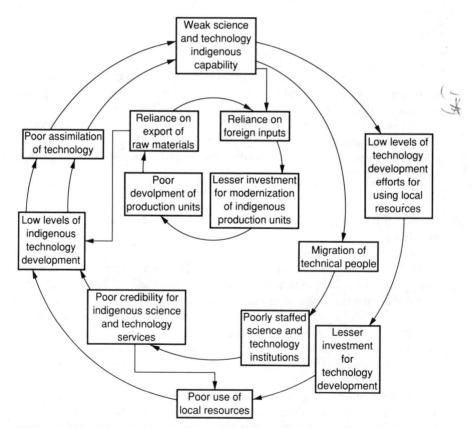

FIGURE 1.1 The vicious circles of lack of technology and underdevelopment. (Reprinted with permission from *Technology Atlas Project*, UN-ESCAP/APCTT.)

human mind to remember and perform both simple and difficult operations. Emerging technologies are clearly identifiable sources of macro-socialization because these technologies have provided both the channels for information flow and the tools for information processing. Computers, for example, have for several decades served as tools for supplementing the memories and limited cognitive powers of decision makers. Products obtained through technological innovation have demonstrated the capability to enhance the quality, effectiveness, and productivity of a decision-making behavior or a decision-making process. This notion is also supported by the Technology Atlas Team (1987), which stated that technology, in today's increasingly interdependent society, provides hope, values, and faith for mankind. It brings hope for bridging the gap between the haves and have-nots; it is responsible for altering economic and social values; and it is the faith upon which the world of tomorrow is being built.

If the vicious circles of lack of technology and underdevelopment shown in Figure 1.1. are to be broken, it is necessary to understand the problems, constraints, and opportunities in achieving technology-induced progress. At the enterprise level, technology is even more critical because firms are characterized by rapid changes in competitive market strength and are continually faced with problems such as product life cycles, technology gaps, optimization of resources, and global competition. Technology is, therefore, a competitive weapon at all levels of the economy.

It must be pointed out that useful technologies and their effective management come from an appropriate innovation and technology management process that has in place adequate techniques and methodologies for comprehensive planning and forecasting, innovation development, technology assessment, measurement, and control.

1.2 DEFINITION OF INVENTION AND INNOVATION

1.2.1 Invention

Invention can be defined as the creation of a new idea for a product, process, or service. Invention is a new combination of preexisting knowledge that satisfies some want. When an enterprise produces a new good or service or uses a new method or input, the enterprise makes a technical change. The first enterprise to make a given technical change can be viewed as an inventor and its action as invention.

1.2.2 Innovation

Innovation is the introduction of a new product, process, or service into the marketplace. Innovation is not a technical term; Drucker (1974, p. 785) states that innovation is an economic and social term. Its criterion is not science or technology, but a change in the economic or social environment, a change in the behavior of people as consumers or producers, as citizens, as students or teachers, and so on. Innovation creates new wealth or new potential of action rather than new knowledge. Therefore, the bulk of innovative efforts will have to come from the places that control manpower and money needed for development and marketing, that is, from the existing large aggregation of trained manpower and disposable money—existing businesses and existing public-service institutions. The sources of opportunity for innovation suggested by Drucker (1985a, p. 68) are:

1. Unexpected occurrences.
2. Incongruities.
3. Process needs.
4. Industry and market changes.
5. Demographic changes.
6. Changes in perception.
7. New knowledge.

Marquis (1969) defined the types of innovations as follows:

- Radical innovations: ideas that have impact on or cause significant changes in the whole industry.
- Incremental innovations: small ideas that have importance in terms of improving products, processes, and services.
- System innovations: ideas that require several resources and many labor-years to accomplish. Communications networks and satellite operations are good examples of system innovations.

Innovation Interfunctional Process. Innovation may be viewed as an interfunctional process consisting of three major functional areas:

1. Research and development (R&D), which creates.
2. Manufacutring, which produces.
3. Marketing, which sells.

The R&D function is concerned with basic and applied research and the developmental use of the outcome of such research. The manufacturing process transforms inputs to output based on the set of guidelines presented at the development stage. The marketing function helps take the product to the marketplace. Other functional areas, such as personnel services, finance, information systems, production control, and maintenance, are designed to support these three categories. The process of innovation on an existing product or completely new product can take place in any given functional area.

Betz (1987) pointed out that technological invention is the first step in technological innovation. At any time, technological invention draws on the current

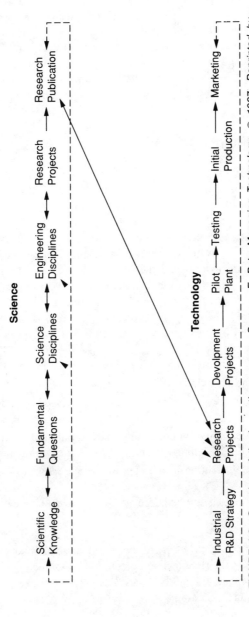

FIGURE 1.2 Overview of the innovation process. Source: F. Betz *Managing Technology*, © 1987, Reprinted by permission of Prentice-Hall, Inc., Englewood Cliffs, NJ.

scientific knowledge base. An overview of the innovation process presented by Betz is shown in Figure 1.2.

Four Major Features Evident in a Technologically Innovative Environment. In most advanced economies four features have been evident: (1) dynamic management approaches to labor, money, materials, and other natural resources; (2) the need to be innovative in order to be competitive; (3) the use of imaginative financing methods to provide the financial resources for innovative projects; and (4) a strong awareness that the rapid growth and obsolescence of technology call for an ongoing innovative process in a dynamic environment.

1.3 INNOVATION PROCESS MODEL (IPM)

The Innovation Process Model (IPM) presented in Figure 1.3 shows that innovation begins with a new idea that is influenced by some event in either the external or internal operating environment. The operating environment events stimulate the memory, intelligence, and experience of the innovator, and this stimulation leads to the recognition and formulation of both the technical feasibility and the demand for a new idea. The innovator then embarks on problem solving, data gathering, and data manipulation to translate the idea to the form of an invention. In the development and testing stage, the innovator tests two things: the feasibility of the idea in production and the acceptability of the end product in the marketplace. The final stage in the IPM process involves the diffusion of the end product in the marketplace. IPM follows a logical reasoning process that includes:

- Logical organization of a basic idea into meaningful experience.
- Refining the idea for clarity.
- Solving potential problems related to the idea and searching for feasible solutions.
- Revising the idea based on constraints, new input, and other suggestions and embarking on full development and testing of idea components.
- Full-scale implementation of the idea in the marketplace.

Three levels of actions are required to support the elements presented in the IPM.

Level One: Policy Formulation At both the national and firm levels, it is important to formulate the right technological policy for creating or setting the stage for radical, incremental, and systems innovation. Specific areas of technological policy that require attention are:

1. Science and technology education and training.
2. Research and development organization and guiding policies.

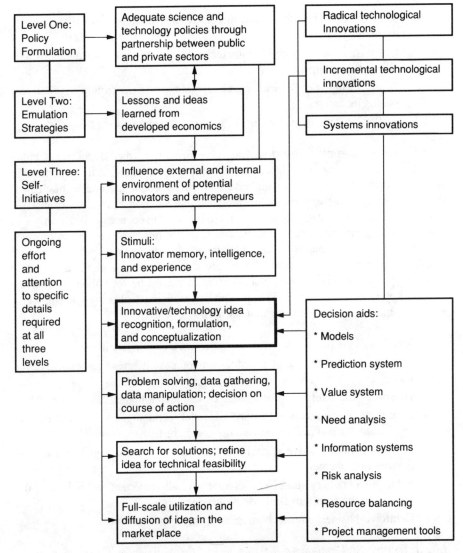

FIGURE 1.3 Innovation Process Model (IPM).

3. Research and development funding to enhance science and technology development.
4. Mechanisms for local technology development, manufacturing, and marketing.
5. Mechanisms for selecting new technologies.
6. External trade balance and how it affects sources of technological capability.

7. Incentives that provide tax advantages, loans, or subsidies for technology programs.
8. Mechanisms for encouraging joint technology ventures between the business units of the enterprise.
9. Manufacturing, development, and marketing process technology delivery systems incentives.
10. Developing sources of technical ideas and information exchange.

Level Two: Emulation Strategies Emerging firms have the advantage of being able to learn from the experiences of developed companies. After World War II, for example, Japanese firms embarked on strong economic and technological development using the experience, techniques, and technological ideas from U.S. companies. In the 1980s, Japan has not only shown economic leadership but maintained a competitive edge in some key technologies. The emulation process, if employed skillfully, has enormous advantages. It reduces the capital and overall resource expenditures that would be involved if the technological idea had to be developed from scratch. Emulation strategies should be formulated during the planning process. Porter's (1985) approach for planning technology strategy, for example, incorporates emulation strategy well.

1. Identify all the distinct technologies and sub-technologies in the value chain.
2. Identify potential relevant technologies in other industries or scientific development.
3. Determine the likely path of change of key technologies.
4. Determine which technologies and potential technological changes are most significant for competitive advantage and industry structure.
5. Assess a firm's relative capabilities in important technologies and cost of making improvements.
6. Select a technology strategy, encompassing all important technologies, that reinforces the firm's overall competitive strategy.
7. Reinforce business unit technologies at the corporate levels.

Level Three: Self-Initiatives At this level, the policies and strategies of the first two levels influence individual person, firm, or organization. Influences from the external and internal environment also help stimulate innovator memory, intelligence, and experience. Self-initiatives for technological development and growth can be encouraged through

1. Adequate facilities and infrastructure for science and technology projects.
2. Systems and structures that promote individual and organizational effectiveness and efficiency.
3. Development of individual capabilities, including foresight, creativity, balanced judgment, self-confidence, endurance, and interpersonal skills.

4. Unstructured and structured environments that permit and encourage innovation.
5. Adequate management system for people, process, and projects.

Level three is also concerned with the process of evaluating technological alernatives, which should consider the following items:

1. The range of the new technology's potential applications in the external and internal market.
2. The nature of the external competition and ways to satisfy market demand.
3. The skill required to develop and implement the new technology.
4. Opportunities for employment or unemployment that the new technology will create.
5. The sources of raw materials and other resources required to produce the new technology.
6. The optimal balance of performance, feature, efficiency, and costs involved to position for market.
7. The infrastructure required for the new technology.
8. The management system required for the technology project, modes of distribution, and marketing approaches.
9. Impact of regulations and other environmental, political, and social constraints.
10. Technology licensing, timing of entry, and research and development expenses required.

Selecting appropriate technological alternatives can also be aided by having the right group of technical personnel—with experience in marketing engineering, manufacturing and development, business management, and research issues—

TABLE 1.1 Technology Selection Decision Matrix

Technology Options	Section Impact[a]				Risk Involved		Investment and Time Required		Constraints		Market Potential		Decision	
	A	M	D	S	High	Low	High	Low	High	Low	High	Low	No Go	Go
High technology	x				x		x		x		x			x
Medium-range technology		x	x		x		x				x	x		
Low-range technology	x				x		x			x	x			x

[a] A = agriculture, M = manufacturing, D = defense, S = service.

involved in the planning process. Table 1.1 presents a technology selection decision matrix for use in evaluating technological alternatives and selecting the most appropriate technology for development. Techniques for justifying new technologies are discussed in Chapter 4.

1.4 TECHNOLOGY TRENDS

Technology is a specialized body of knowledge that can be applied to achieve a mission or purpose. The knowledge can be in the form of methods, processes, techniques, tools, machines, materials, or procedures. Technology can therefore be defined as the means by which knowledge is applied to produce goods and services. Some of the earlier technologies were stones for creating a fire, bow and arrow for hunting, and canoes for navigation. In modern times new technologies such as computers, robotics, expert systems, and lasers have emerged. But technology is not limited to inanimate objects alone. Even during modern times, animals such as dogs, elephants, and dolphins are used to provide services that would otherwise require automated technology.

Trends in technology that have become important for service and manufacturing operations include information-processing technology, computer-aided design (CAD), computer-aided manufacturing (CAM), computer-integrated manufacturing (CIM), group technology, robotics, automated assembly technologies, lasers, microelectronics, telecommunications technologies, and process technology. High-technology applications in manufacturing and service work environments are estimated to grow by 35% or more annually. Figure 1.4 shows the rapid growth in number of computers in the United States alone.

The Electronic Industries Association (EIA) reported that 138,000 minicomputers, defined as those priced between $5,000 and $40,000, were sold in the United States in 1979 alone. Purchases of the compter-aided design (CAD) systems, the first step in integrated computer-aided manufacturing systems (CAD/CAM), amounted to about $570 million in 1979. CAD purchases could hit $9 billion by the year 2000. In 1982, an International Data Corporation study estimated that there were about 730,000 computer systems in operation in the United States. In 1981, the Robot Institute of America conducted a survey and estimated a total of 4,100 robotic devices in use in the United States compared to 67,435 in Japan. An item in *Industrial Engineering*'s new front section in July 1983 predicted an increase in industrial robotic sales from about 1,700 units in 1982 to 38,000 annually by 1992, with annual sales to exceed $2 billion by 1992.

From 1948 to 1958, technological change in electronics progressed from the invention of the transistor, the development of the germanium transistor chip, the invention and development of the silicon transistor chip, and the invention of the integrated circuit (IC) of silicon semi conducting chips (Betz, 1987). From 1960 to 1980, progress in semiconductor chips continued linearly as transistor density on chips was increased.

Two major advances in the early 1980s increased transistor density experimentally: large-scale integration (LSI) and very large-scale integration (VLSI). There

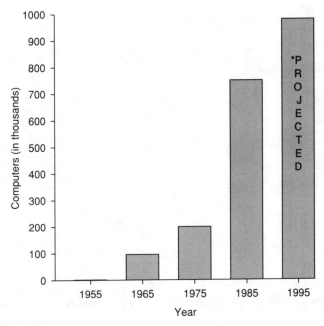

FIGURE 1.4 Estimated number of computers in use in the United States.

is every reason to expect that the number of transistors per chip could climb to a number ranging from several hundred million to about one billion by the year 2000. Experts describe the next era as ultra large-scale integration (ULSI). The improved computer performance shown in Figure 1.5 can be attributed to significant changes in electronic devices and overall improvement in total system logistics. Although more attention has been given to computer and robotic devices, other technologies such as lasers, ultra-sound satellites, and machine tool applications have also increased in large numbers in the last several decades. There is no doubt that technology applications in both the manufacturing and service environments will continue to increase in large numbers. Our expanded information society continues to call for more technology in the decision-making process.

As the technology explosion continues, both managers and employees will have to become more familiar with data-intensive techniques. Goldhar and Avakian (1979) note that data-intensive production has two important characteristics that complicate its management: inalienability and indivisibility. *Inalienability* means that information is not bound to only one user—new users and the original owner can possess it at the same time at little or no added cost. This is called economy of scope. *Indivisibility* means that one cannot transfer or use partial information; all of it is required for desired outcomes. Inalienability and indivisibility lead inexorably to an emphasis on compatability and standardization.

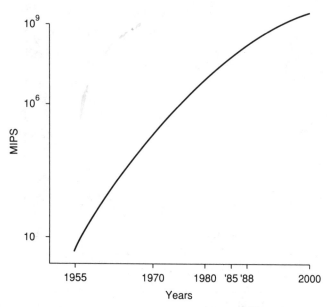

FIGURE 1.5 Computer Performance (MIPS = million instructions per second).

1.5 IMPACT OF TECHNOLOGICAL PROGRESS

Technological progress is the result obtained when technology is used in an effective and efficient manner to improve productivity, reduce waste, improve human satisfaction, or enhance the quality of working life.

Technological progress is one of the major forces driving industrial growth. It enables more output to be produced for a limited use of input resources. Between 1870 and 1957, 90% of the rise in real output per labor-hour in the United States can be attributed to advances in technology. (Abramovitz, 1956; Kendrick, 1961) There is a consensus among experts that about 40% of the total increase per capita income is due to the impact of technological change. Technological progress is not, however, a phenomenon that stands by itself; it is driven by effective human resources and effective management systems. In turn, technological change affects the quality of working life at different levels.

Social Aspects Woodward (1958, 1965, 1970), Blauner (1964), Faunce (1968), and Whyte (1961) discuss at length the social consequences of new technology applications. Research findings by these and other experts suggest that new technology has had a pervasive influence on the quality of working life. People have become separated from work in varying degrees according to the kind of work they do and their position within the organizational hierarchy. This separation has led to varying degrees of alienation, boredom, and apathy at work, along with loss of selfhood. Several studies have examined the impact of computer applications in the work environment. (Mann and Williams, 1960; Hoos, 1961; Lee, 1965a and 1965b; Whisler, 1970; Bjorn-Anderson, 1976; Bradley, 1977; Norman, 1980; Turner, 1980; Karasek and Turner, 1981; Edosomwan, 1985a, 1985b, and 1985c). They have found the introduction and use of the computer-aided method in routine assembly tasks to have the following effects:

1. Make the job more productive, in that output per worker (labor productivity) is greater.
2. Make the job less autonomous in that control is taken away from workers.
3. Make the job more demanding in that the pace of the job and the workload increase.
4. Make the job more formalized in that there are more rules and procedures to follow (increased complexity).
5. Make the job more anxiety-provoking in that workers report greater stress and strain on the job from increased workload.
6. Have impact on employment in that there is a reduction in the number of workers required to do a given job after computerization.

Macroeconomic Aspects Skinner and Chakraborty (1982, pp. 6–7) summarize other impacts of technology as follows:

New technology and automation of functions once formed by humans create employment problems. Briefly, these problems arise due to plant shutdowns, the layoff of workers with or without shutdowns, and the mechanisms of adjustment to this change. Remedial measures involve both public and private efforts to mitigate the harsher effects of technological change on the work force.

Skinner and Chakraborty further suggest that technological change may be more broadly based than its impact on the immediately affected industry. They cite as an example the favorable effect of technology that reduced price and time factors in the travel industry had on hotels, motels, and resorts.

Educational Aspects Skinner and Chakraborty (1982) and Jahn (1973) point out that technological changes have created job requirements that differ from the old ones. As a market response, education tends to prepare people for these new jobs within a curriculum that reflects this change. Berger (1964) suggests that the changing nature of education creates new professions, which must establish themselves over time. Acceptance is difficult at the beginning, but with time, new professions tend to legitimize themselves through professional societies and organizations. Technology management, for example, has become a growing field of education and research, and I predict that by the year 2000, several higher education institutions in business and engineering will be offering courses and degrees in this field. Edosomwan and Khalil (1988) organized the first international

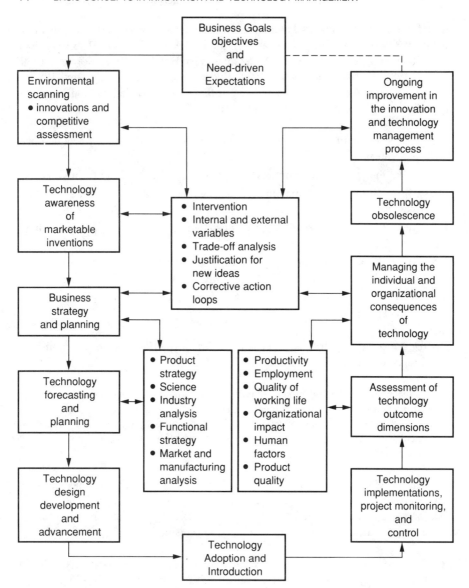

FIGURE 1.6 Elements of the innovation and technology management process.

conference on technology management. The conference, which was held in Miami, Florida, discussed along with other subjects curriculum for technology management education. Several experts from more than 35 countries agreed that the management of technology should be treated as a discipline. Brief discussions of other challenges involved in technology management are presented in Section 1.9.

1.6 DEFINITION OF INNOVATION AND TECHNOLOGY MANAGEMENT

Innovation and technology management requires an integrated process involving both management and employees with the ultimate goal of managing the invention, design, development, production, transfer, introduction, and use of the various forms of technology in the work environment. Figure 1.6 presents the key elements that must be continuously managed in the innovation and technology management process. Useful technologies and their effective management are achieved through the innovation and technology process if there are techniques and methodologies in place for comprehensive technological planning, forecasting, assessment, introduction, measurement, control, evaluation, implementation, improvement, and monitoring. An effective systems approach that involves both management and employees in bringing forth new innovations and technologies is recommended, as is the management of both positive and negative consequences of technology.

1.7 MANAGEMENT OF TECHNOLOGY (MOT)

The Task Force on Management of Technology defines management of technology (MOT) as follows:

> Management of technology is an industrial activity and an emerging field of education and research that is not generally well recognized or even consistently defined. It concerns the process of managing technology development, implementation, and diffusion in industrial or governmental organizations. In addition to managing the innovation process through R&D, it includes managing the introduction and use of technology in products, in manufacturing processes, and in other corporate functions.

According to the Task Force on Management of Technology (1987):

> Management of technology links engineering, science, and management disciplines to plan, develop, and implement technological capabilities to shape and accomplish the strategic and operational objectives of an organization.

Key elements of MOT in industrial practice are (1) the identification and evaluation of technological options; (2) management of R&D itself, including determination of project feasibility; (3) integration of technology into the company's overall operations; (4) implementation of new technologies in a product and/or process; and (5) management of obsolescence and replacement. The primary industry needs identified by the MOT Task Force are:

1. How to integrate technology into the overall strategic objectives of the firm.
2. How to get into and out of technologies faster and more efficiently.
3. How to assess/evaluate technology more effectively.

4. How to best accomplish technology transfer.
5. How to reduce new product development time.
6. How to manage large, complex, and interdisciplinary or interorganizational projects/systems.
7. How to manage the organization's internal use of technology.
8. How to leverage the effectiveness of technical professionals.

Edosomwan et al. (1988) points out that management of technology at the business enterprise level in many different ways, none of them necessarily right or wrong. Differences usually occur regarding the scope of the elements composing technology management. The following text focuses on four major aspects for managing technology: integrating processes, managing the product cycle from concept to commercialization, leveraging resources, and meeting business goals and objectives. Managing technology can be defined as a method of operation that leverages human resources, technology, and other business assets by optimizing the relationships between the technology functions of the business enterprise. It is the process of integrating science, engineering, and management with research, product development, and manufacturing in order to meet the operational goals and objectives of the business enterprise effectively, efficiently, and economically. It includes managing the totality of the technology operations from product concept to commercialization.

Managing technology deals with many issues that affect firms' competitiveness in the global market. Its scope includes such topics as technology transfer, development, implementation, and diffusion in industrial and service organizations. It includes approaches to managing (1) innovation, (2) entrepreneurship within firms, and (3) the introduction and use of new technology in products, processes, and other organizational functions. More specifically, managing technology

- Integrates science, engineering, and management with research, product development, and manufacturing.
- Emphasizes managing total performance in the entire cycle from product concepts to commercialization.
- Focuses on activities that reduce total cycle time, contribute to improved business performance through "productivity improvement" and improve responsiveness to customer needs.

1.7.1 Managing Technological Changes and Discontinuities

Sumanth (1988) points out that whenever changes occur in product technologies or process technologies, enterprises face severe challenges, particularly when the competitive forces are quick, unpredictable, and dynamic. The side effects of choosing inappropriate technologies can be significant, as pointed out by Rybczynski (1983). Michael Porter's 1985 work on "competitive strategy" became the hottest topic in the early 1980s when many American automotive companies were facing severe competition from the Japanese. Buffa (1984) shows how new manufacturing strategies must become a vital part of overall corporate strategy for competitiveness and recommends six specific strategies to accomplish this. He addresses the basic elements of manufacturing and the impact of new technologies on productivity. Riggs (1983) also focuses on recommending management strategies to improve competitiveness, with particular emphasis on high-technology companies. More recent works of Quinn (1985), Waterman, Jr. (1987), and Peters (1987) emphasize the growing concern of competitiveness and the need for "revolutionary" management thinking to cope with unusual times of complexity and uncertainty in various areas of management activities, including technological innovation. The intensity of discussion about the rapid pace of product innovation and development abounds in many other contemporary works, including a global thrust by Ohmae (1985).

According to Sumanth (1988), today's technologies are changing very rapidly, and the challenge becomes even greater as enterprises have to keep pace with such technologies at the product and production process level. Consider, for example, the personal computer. Just about 10 years ago, when the first personal computer (TR-80 by Tandy Corp.) was marketed, the product technology offered a capability of an 8K random access memory (RAM). Today, an IBM PS/2 Model 80 can come with as much as 16,000K RAM (2000 times more) with a maximum disk storage of 230 MBytes. Today's surface mounted technology has drastically changed product design considerations compared to even three years ago. The IBM PC XT model, which was introduced in 1986, is no longer produced as the new PS/2 systems come into the market. The average shelf-life of personal computers is now about two years. Imagine the dynamics of change for a typical manufacturer of personal computers. By the time the product is designed, pilot-tested, sourced out to the vendors for parts, assembled, and marketed with a decent advertising program, the competition is already in the "cloning mode," ready to threaten the new product entries. Global satellite communications have made it possible to access vital information on a worldwide basis at an unprecedented speed. Gone are the days when the computer manufacturers can think that they can make all their money in the first three or four years, because by that time, the product itself might become obsolete or not "prestigious" enough to be accepted by customers as state-of-the-art.

Recent works by the New York Stock Exchange (1984), Betz (1987), and Utterback (1986) point out the importance of innovation in becoming competitive, but the nature of change that must take place to rapidly improve innovation is not addressed satisfactorily. Schonberger (1987) proposes "incremental improvements," but they are neither appropriate nor desirable during the periods of technology discontinuities. Monger (1988) strongly points out the lack of management of technology and identifies three problems characterizing the current status of American technology management practices: slow technological absorption, height implementation failure, and avoidance of social consequences.

1.8 S-CURVES

S-curves are obtained by plotting the effort expended to improve a product or process against the performance for that effort. Benefits obtained at the early stages of product development are small, but as additional knowledge and investment are applied, positive results progress rapidly. After an optimum level of effort is applied, progress becomes more difficult. Foster (1986) points out that relatively small gains in technical progress in comparison to incremental increases in the effort for a given type of technology result in a certain S-curve. Technological discontinuities may also exist as one group of products or processes is supplanted by another. Figure 1.7 shows discontinuities occurring when one aircraft propulsion technology replaces another. Emerging firms can usually take advantage of the inherent characterisitics of the technology change process by leap-frogging on the development of early starters and can therefore skip the intermediate stages of technological change. Eventual catch-up with the initial designer and developers of technology is also possible depending on the S-shaped growth pattern of any particular technology. The Technology Atlas Team (1987) points out that late starters have the opportunity for avoiding known pitfalls by:

- Learning effectively from the experiences of developed and other developing countries.
- Preventing or minimizing the possible negative impacts of certain available technologies.
- Carefully studying the successes and failures of one's own past experience for determining future activities.

Drucker (1967) notes that discontinuities occur more frequently than most of us realize and that their frequency is increasing. Choices are often difficult for those caught up in technology discontinuities. Peters (1987) states that management of technological discontinuities requires specifying and measuring technological performance, output as well as input, and seeking and understanding alternative approaches and their limitations. The implications of rapid technology discontinuities in the case of either products or processes are enormous. Sumanth (1988) specifically states at least five major implications:

1. Product Life Cycles. Product life cycles are directly affected by technology discontinuities, thereby making critical the decisions regarding advertising, marketing, out-sourcing, capital equipment purchases, facilities layout, manpower allocation, and so on. There is a "wait-and-see" policy often followed by enterprises as well as customers when there is a risk of rapid obsolescence due to technology discontinuities. Mistakes made by an enterprise in the timing of the decisions can sometimes be catastrophic, with billions of dollars at stake.
2a. Employee Training and/or Retraining. When there are significant changes in product or process technologies, the need for training and/or retraining becomes a significant challenge, particularly when the economic factors

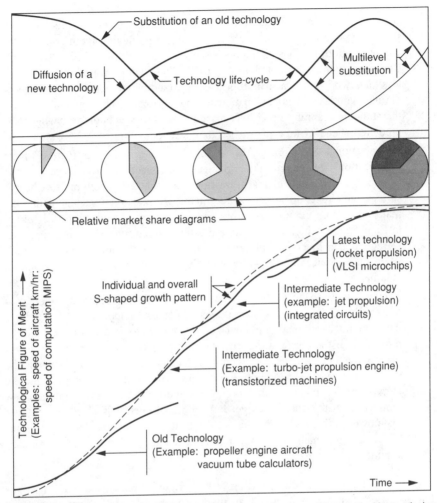

FIGURE 1.7 Sequented S-shaped technological change process. (Reprinted with permission from Technology Atlas Project, UN-ESCAP/APCTT.)

to compete in the market outweigh the social implications. Many com[pa]nies' first line of thinking is to lay off employees, if their corpo[rate] is not emphatically oriented toward "full-employment."

2b. Re-orientation or Re-education of Customers. [...] the chairmanship of the Chrysler Corp[oration...] and process changes, for example, the [...] use of robotics in the welding and painti[ng...] have introduced new product and process[...] customers and retain old ones, he launched [...] "New Chrysler Corporation" image. It wor[ked...]

3. Corporate Strategic Planning. Whenever te[chnological change takes] place, the "shock" on the enterprise can be min[imized...]

are part of a corporate strategic planning process. Of course, prediction of such discontinuities is not at all easy, but not predicting them could put a company out of busines. One good example is that of the Osborne Computer Company, which did not time its product innovation properly and, within less than two years, was bankrupt. Notwithstanding the compounding effects of uncertainties in global, national, and regional environmental factors, the strategic planning process is more robust when technology changes are predicted ahead of time than when they are not.

4. Organizational Structures. Almost always, enterprises "restructure" their organizations whenever major changes take place in product or process technologies. If such changes are not well planned, "cultural flexibility" suffers, which sets up negative forces in the organizational management. One major U.S. corporation went from conventional communication technology to fiber optics, which resulted in a drastic restructuring of the organization. Deregulated airline companies and banks went through similar reorganization when their services had to streamline through better and newer equipment, automation, and so on.

5. Vendor's Inability to Cope. While large corporations can adapt their production systems to rapidly changing technologies, vendors often do not have the economic, technical, and managerial strengths to do the same and still be profitable. If the corporations choose to have "vertical integration" because of employment policies and economic factors of production, they put the vendors in a very vulnerable position. Furthermore, if the external economic factors are favorable to out-source parts and services to the vendors a few months later, the companies get left with too much capital equipment and manpower. Whichever way one looks at this phenomenon, the implications of technology changes are not pleasant for either the corporations or their vendors.

A management guru, Peter Drucker (1969), rightly pointed out that "discontinuities do occur more frequently than most of us realize and, if anything, their frequency is on the increase." So the choices are often difficult for those caught up in technology discontinuities. Often there are many solutions but, quite frankly, no one knows for sure anymore why companies are doing what they are in coping with such discontinuities. This book is not going to offer solutions any easier or better than others but, rather, attempts to expose the reader to a somewhat nontraditional approach to managing such discontinuities, hoping that the reader will adapt the approach appropriately.

TECHNOLOGY MANAGEMENT CHALLENGES

Managing Technology's Impact on Management

to Skinner and Chakraborty (1982), management functions tend to the general as opposed to the specific, as technology changes.

They suggest that as technology changes, more flexibility in managerial training and background will be required. Coordination functions become more important because the number of staff specialists of varying kinds tends to increase. Skinner (1978, 1979) has examined the impact of technology on managers in manufacturing and has found that managers may be averse to handling the consequences of technology directly because they must delegate technological decisions to experts, based on the belief that only years of training can give one the competence required to make high-technology decisions. But Skinner also suggests that successful delegation is, in fact, the key to success. The problem is compounded because many managers cannot conceptualize technology-dependent issues well. What the manager needs to do is acquire a general understanding of technology.

Colding et al. (1977) report that design is one area in which technological developments will affect management. Miller and Armstrong (1966) examined the effects on management structure at a steel plant undergoing conversion to computer-based automatic production processes. They found that the social and economic factors that the firm had to deal with were as important as the technology in determining the structure of the management function.

It is evident that as technology changes, the nature and activities of the managerial function will change. Because technological change in both the manufacturing and service work environments will be ever-present in a competitive world economy, management must learn and find innovative approaches to cope successfully with it.

1.9.2 Managing Technology's Impact on Workers and Professionals

Bechofer (1973) stated that technology affects the well-being of the worker through its effects on the structure of work. Walker and Guest (1952) also point out that production technology directly affects not only the job but also the nature of the worker's in-plant anxiety, affecting relationships with both co-workers and management. The authors suggest that, despite expressed satisfaction regarding rate of pay and security, the average worker appeared oppressed by anonymity. Technological change also brings job dislocation, unwanted relocation, layoffs, and loss in pay, promotion, and seniority. Bok (1964) points out that as technology changes, attrition is the leading method of avoiding layoffs. The challenges that workers face as new technologies are introduced has been summed up by Kheel (1966): "The general attitude toward the effects of technological change—spare the worker but not the job."

1.9.3 Managing the Shrinking Technology Life Cycle

In the early 1960s the average life cycle of technologies was about 10 years. In 1986 most technology life cycles had been shortened to about 2 years, and are projected to be six months by the year 2000. The constantly shrinking average life cycle of a typical technology is presented in Figure 1.8. The dynamic changes in technology life cycle call for new methodologies and techniques, technology

22 BASIC CONCEPTS IN INNOVATION AND TECHNOLOGY MANAGEMENT

feasibility studies, specification monitoring, model development, prototype testing, product manufacturability, and serviceability. The analysis and understanding of the input and output components in each technology life-cycle stage are crucial to the successful applications of the new technologies.

1.9.4 Providing Communications Ability for Technologies under One System Network

A major problem that exists today in technology applications is that the various computer systems cannot communicate with each other under one network. The ability to do so will become an absolute necessity in an information-based society.

1.9.5 Managing the Human Total Reliance on Technology

Now, and in the years to come, managing the human total reliance on technology when the system reliability is less than 100% is a major challenge. Workers and decision makers often become so dependent on the use of technologies that crises develop when the system is down. The consequence is usually a decline in labor and capital productivity. For example, if the large computer system supporting four of the largest banks in the United States is down for 24 hours, the U.S. gross

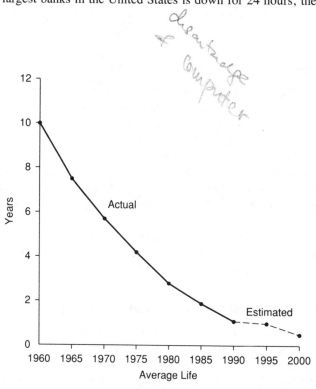

FIGURE 1.8 Average life-cycle trend of a typical technology.

national product will be seriously affected. There is need for ongoing development of error-free systems that can run as stand-alone units.

1.9.6 Designing Acceptable Technologies

Improvement in productivity is the most common rationale for the development and use of new technologies. Experiences from the last four decades have shown, however, that a comprehensive approach recognizing both human and production-related variables is needed for designing new technologies. Chapter 3 discusses the comprehensive design rules for emerging technologies. System designers should consider human anatomical structures and capabilities, morale issues, job satisfaction issues, quality, productivity, and systems flexibility in all stages of the technology life cycle.

1.9.7 Managing the Introduction of New Technologies into the Work Environment

Due to the increasing numbers of new technologies expected in the work environment, decision makers (engineers in particular) will be faced with the complex problem of how to introduce a new technology into the workplace. A three-step approach should be used: (1) perform technology task environment analysis to understand the implications from a user's perspective; (2) provide training facilities for personnel using technology; and (3) involve the user in the maintenance and adaptation issues that are likely to occur in the work environment. A more detailed methodology for implementing technology in the workplace is presented in Chapter 4.

1.9.8 Managing the Impact of Technology on the Quality of Working Life

Before and after the introduction of a new technology into the work environment, it is important to understand the implications for the quality of working life. Most technology, once introduced to the work environment, can cause lowering of job satisfaction, employee morale, and occupational safety.

1.9.9 Managing Technology Systems Requirements

Emerging technologies still lack ability for voice recognition and synthesis, ability to understand natural languages such as English, and ability for automatic programming.

1.9.10 Managing Information

In recent years, the number of computer crimes related to information theft has been on the rise. Techniques are needed for information asset security, and new tools are needed for managing the large volume of information generated by the massive application of the various forms of technologies. New techniques are also needed to minimize system errors, improve system reliability, and reduce the high cost of system requirements. Adequate response time and flexible system-sharing options are still needed for most operating systems.

1.9.11 Managing the Impact of Technology on Productivity and Quality

Understanding the true total cost of doing business in a high-technology work environment will become increasingly important in the years to come. New techniques for assessing the input and output relationship in the technology life-cycle stages are required. The quality of the output obtained from new technologies must also be assessed. Chapter 4 discusses new techniques for productivity assessment in a high-technology environment.

1.9.12 Managing Workers in a High-Technology Environment

Although several motivational techniques exist today, the effectiveness of the existing techniques may decline in the era of expanded technological change expected in the years to come. New motivational techniques that address the changing values and roles of workers as technology changes are needed. Flexible working conditions and organizational structures may have to be adopted to accommodate the changing values of men and women in a technologically advanced society. Technology will probably make it possible for many to work at home, causing performance planning evaluation to become more difficult.

1.10 Understanding Technology Outcome Dimensions

It is important to understand the various dimensions associated with the use of technology such as computers and robotics in the work environment. To do this, the Technology-Aided Task Impact Model (TATIM) shown in Figure 1.9 highlights the production output variables and the human-related variables that are likely to be affected when technology-based systems are used in the work environment.

In situations where technology-based systems are used to assist humans in producing goods and services, the work environment can be described as one where task structure mediates between the information system and the worker. The worker's job satisfaction and psychological stress are therefore affected by changes in the task structure associated with changes in the system design parameters. Other outcome dimensions such as productivity and quality are affected by changes in the mechanism of production or service. Employment levels are affected by changes in productivity. Occupational safety and working conditions, system outcomes, and ergonomic factors are affected by system and task design parameters. In a work environment that uses technology, the optimization of TATIM variables and parameters requires an ongoing process of planning, measurement, evaluation, control, improvement, and maintenance.

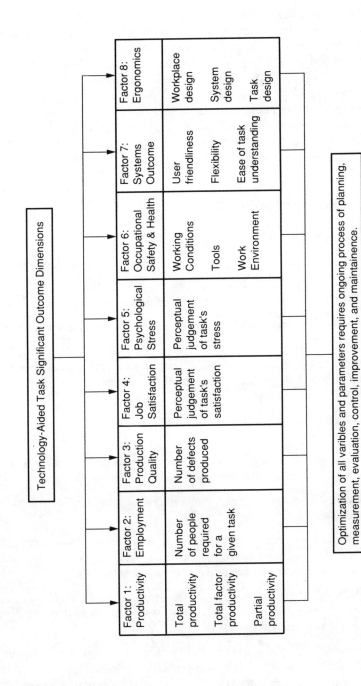

FIGURE 1.9 TATIM Outcome Dimensions

SUMMARY

The intent of this chapter has been to present some key definitions and concepts in innovation and technology management. There is no doubt that technology will continue to provide humankind with the capability to process quantities of information and solve problems that would otherwise be impossible to handle. More technological changes are expected to have more impact on the structure and functions of the work organization and the quality of working life. A thorough understanding of the innovation process and the outcome dimensions associated with the use of the various technologies provide a starting point for decision makers. In the innovation and technology management process, the decision maker's role becomes that of a system integrator.

QUESTIONS AND PROBLEMS

1-1 Define technological innovation and technology management.

1-2 Present a solution strategy to help decision makers address the innovation and technology management challenges of the year 2000.

1-3 How can an organization protect itself against computer crime?

1-4 Define the various types of innovation and present an innovation model for a new robotic device.

1-5 Assume you are the chief executive officer of a major corporation involved in high technology. How would you formulate a technological policy for your corporation?

1-6 Suggest a procedure for managing the shrinking change in the technology life cycle.

1-7 What are the impacts of technological innovation on competition, productivity, and the quality of working life?

1-8 Suggest a solution strategy to manage the negative impact of technology on the quality of working life.

1-9 Describe the role of a project manager in an information-based organization. Suggest training requirements for managers, engineers, and analysts in handling programmed and nonprogrammed information.

1-10 How would you go about motivating workers in a high-technology environment? What type of leadership style would you adopt in managing a group of talented innovators?

1-11 Assume you are the chief executive officer of a high-technology company. How would you proceed in implementing the management of technology recommendations offered by the National Academy of Science Task Force (1987)?

1-12 Develop a technique for integrating technology into the overall strategic objectives of a firm of your choice.

1-13 What are the mechanisms that can enable designers and developers of technology to get product transfer from one location to another?

1-14 Develop a strategy for evaluating a new technology implementation in the work environment.

1-15 How should technical professionals be managed in a high-technology environment?

1-16 How should product design, development, manufacturing, and marketing time be shortened? What is the impact of a product's life cycle on the product's final price?

1-17 What are the requirements for successful management of technology?

1-18 How can large corporations that invest heavily in research and development of new technologies protect their assets from emulation by late starters?

1-19 How would you go about minimizing the potential impact of new technology on workers, management functions, and employment?

1-20 Suggest a procedure for managing technology gradient in a high technology environment.

1-21 Does technological progress bring about improvement in the quality of working life? Why or why not?

1-22 What improvement strategies and policies should be implemented at the national level to avoid the vicious circle of technological underdevelopment?

1-23 What techniques should be implemented for information asset security at the company level?

2
REQUIREMENTS FOR INNOVATION AND TECHNOLOGY MANAGEMENT

This chapter discusses the essential requirements for the management of technological innovation and technology issues at the firm level. A strategy that will enable decision makers to establish an effective innovation and technology management program is presented. The chapter examines the impact of the operating environment on the innovativeness of individuals and the attributes of the ideal innovative manager. It provides a guideline for successful supervisory performance and discusses strategies for preparing for the next generation of technology.

2.1 MANAGEMENT'S ROLE IN INNOVATION AND TECHNOLOGY ISSUES

Successful technological innovation requires a competent management system. Effective managers are needed to set the right technological policy for the creation of an environment conducive to technological innovation, and to prioritize available resources to facilitate the development, manufacture, and marketing of new ideas and ventures. Kanter (1982) has studied 165 middle managers in five companies and has found that a firm's productivity depends to a great degree on how innovative its middle managers are. Innovative managers tend to be visionary, comfortable with change, and persistent. Innovation flourishes in firms where (1) territories overlap and people have contact across functions, (2) information flows freely, (3) many managers are in open-ended positions, and (4) reward systems look to the future, not the past.

Twiss (1980) also concludes that top management plays a critical role in creating organizational attitudes favorable to innovation. Twiss further states that technological innovation is intimately related to the management of organizational change. The importance an industrial organization attaches to the management of technological innovation must therefore be reflected in the selection, development, and congruence (or "fit") of Research and Development (R&D) personnel not only at all levels of the R&D functional unit but at the corporate level as well. Twiss explains the findings of the National Research Council of Canada, which studied 95 projects, as follows: "All of the successful projects had at least one able dedicated leader pushing them. No project was successful without a person behind it" (p. 16). Twiss suggests that a formal management system is not sufficient in itself but must be associated with the efforts of at least one highly motivated *technological entrepreneur* or *project champion*. Ramo (1980), Betz(1987), and Moranian (1963) also agree that the technological entrepreneur makes technological innovation possible through good leadership traits. The ultimate success of any organization depends on the skill of top decision makers and how effectively they use other available resources. Effective management of technological innovation requires ongoing development and training for managers and an innovative approach to the management of people, customers, and the total organization.

Another viewpoint that advocates an effective system of management for technological innovation is that of Abernathy (1982). In his studies of consumer electronics and auto industries between 1950 and 1960, Abernathy found that there were no deficiencies in U.S. technological capabilities, nor were there structural barriers in the industry to preclude innovation. No direct capital deficiencies stood in the way, nor was inflation a problem. But management decided to take a short-run view and did not innovate competitively. Abernathy points out that the key to stemming the decline in U.S. competitive performance is in the hands of management. A participative management style that encourages new ideas, welcomes change, and rewards innovative individuals and good performance is recommended.

2.2 FORMULATING THE RIGHT TECHNOLOGICAL POLICY

It is important for a firm to set a technological policy that creates or sets the stage for various innovation aspirations, processes, and achievements. Maidique and Patch (1978) present six major areas of technological policy that a firm can focus on:

1. Technology selection, specialization, or embodiment.
2. Improving the level of competence, with an emphasis on basic research, applied research, and development engineering.
3. Sources of technological capability: internal versus external.
4. Research and development investment and staffing.
5. Competitive timing: initiative versus responsiveness.
6. Research and development organization and policies: flexible or structured.

Maidique and Patch further suggest that appropriate strategy depends, in turn, on the capabilities and resources available to the firm, on the competitive opportunities and threats faced by the firm, and on the objectives of the firm. Technology policy is viewed as a portfolio of choices and plans that enables the firm to respond effectively to technological threats and opportunities.

Edosomwan (1986a, 1987a) states that technology strategy should be formulated within the framework of overall business planning. The four major components of business planning recommended include environmental analysis, strategic planning, tactical planning, and operational planning. The elements of technological policy that should be evaluated along with business planning objectives are presented in Figure 2.1.

Maidique and Zirger (1984, p. 201) state that the following circumstances can improve the chances of success.

1. The developing organization, through in-depth understanding of the customers and the marketplace, introduces a high performance-to-cost ratio.

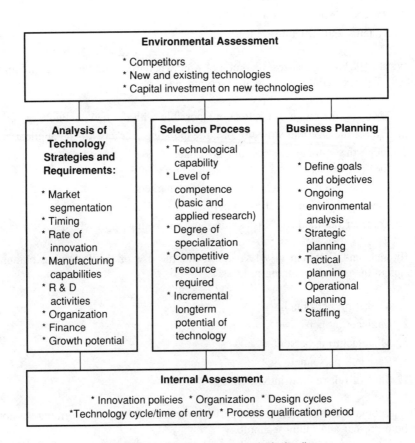

FIGURE 2.1 Elements of technological policy.

2. The developing organization is proficient in marketing and commits a significant amount of its resources to selling and promoting the product.
3. The product provides a high contribution margin to the firm.
4. The R&D process is well planned and executed.
5. The create, make, and market functions are well interfaced and coordinated.
6. The product is introduced into the market early.
7. The markets and technologies of the new product benefit significantly from the existing strengths of the developing business unit.
8. There is a high level of management support for the product through its launch to the marketplace.

Frohman (1982, p. 97) also mentioned that corporate culture can help the chances of success. The three aspects in the corporate culture that are important are

1. Top management orientation.
2. Project selection criteria.
3. Systems and structure.

Ward (1981) emphasizes market position when formulating a technological policy. The six major actions Ward recommends are

1. Focus on the range of applications possible with the new technology.
2. Project the size and structure of corresponding markets for the applications.
3. Judge the optimal balance of performance, features, and costs to position for the markets.
4. Consider alternate ways to satisfy these markets.
5. Analyze the nature of the competition.
6. Consider the modes of distribution and marketing approaches.

Bitondo and Frohman (1981) recommend the use of the following planning groups in the overall planning process:

1. Marketing personnel.
2. Engineering personnel.
3. Manufacturing personnel.
4. Research personnel.
5. Staff of relevant business units.

They recommend that these groups perform strategic R&D brainstorming sessions on both present and future technology.

Maidique and Patch (1978) point out that the technological policy appropriate for an individual firm could depend on whether the firm was adopting a first-to-market, second-to-market, late-to-market, or market segmentation strategy.

2.3 PROMOTING INNOVATION THROUGH COOPERATION AMONG INDUSTRY, GOVERNMENT, AND ACADEMIA

One nation that stands out in fostering good relationships among industry, government, and academia is Japan. These relationships have provided a competitive advantage for Japanese firms, especially in consumer electronics, computers, automobile, camera, and air conditioning industries. The lack of knowledge and understanding in academia about commercialization of technological developments can be quickly overcome through cooperative education programs, joint research between academia and industry, and government support of innovative projects by imaginative and creative financing. There is a strong need for cultivating more understanding among academic people about business practices, especially in the high-technology areas of computers, robotics, and defense systems. Here are some recommended mechanisms for developing and stimulating the field of innovation and technology management.

1. Build awareness through adequate partnership between business units in innovation and technology management matters. Continue to allocate resources to support innovative projects, which are increasingly important to the competitiveness of industry. Resources provided should include but not be limited to:
    - Long-term grants to individual academic or private researchers in innovation and technology management.
    - Funding small-group research, especially joint university-industry efforts.
    - Long-term loans to small businesses seeking breakthroughs in new technologies.
    - Financial support for graduate students specializing in technology management.
    - Conducting a state-of-the-art review of innovation and technology management periodically to assess its progress and directions.
    - Stimulating the interest of other agencies by holding interagency conferences and disseminating the results of such conference through publications.
2. Industry should commit itself to the process in the following ways.
    - Recognize the potential for technological innovation and technology management to be improved through research and education.
    - Work with academia in identifying internal corporate needs, both long-term and immediate.

- Employ graduates well grounded in innovation and technology management and provide them with an opportunity to influence technology-related decisions at levels appropriate to their position in the company.
- Form partnerships with academic innovation and technology management researchers; that is, provide funding for research and curriculum development and make people and facilities available for experimentation.

Management of Technology Task Force Report (1987) recommends that the university community can contribute in the following way:

- University administrators and faculty must recognize the increased importance and validity of problem-oriented research and education in cross-disciplinary fields, and in their deliberations governing promotion, tenure, and reward.
- Tenured professors in related areas could contribute to the rapid development of the field by focusing their research and teaching on needs identified.
- Researchers active in innovation and technology management might benefit from forming a professional association dedicated to the improvement of this field through research and education.

2.4 PROMOTING INNOVATION THROUGH TECHNICAL INFORMATION EXCHANGE

The United States leads the world in the amount of technical information generated and published. This is perhaps one of the reasons why more radical, systematic, and process innovations have taken place United States than in the rest of the developed world. The innovation process is best promoted by an interactive environment where researchers and practitioners can exchange ideas and information. Allen (1978, p. 5) states the importance of interactive association: "The technologist must obtain his information either through the very difficult task of decoding and translating physically encoded information or by relying upon direct personal contact and communication with other technologists." At IBM, Xerox, AT&T, GM, and other major corporations, several benefits have been derived from the cross-fertilization of ideas. The myriad technical papers published yearly in the United States and the rest of the world have also supported many innovative projects.

Although information exchange between practitioners and researchers is highly recommended, it should be done without conflict to the organization's goals or society's values. Interorganizational technical interaction is a complex process with sensitive dynamics. It can bring enormous potential benefits as well as disastrous consequences. Some emerging firms can pave the way to new technological breakthroughs through information exchange, but information exchange channels have also been responsible computer crime and the unlawful use of competitors' assets. Organizations involved in active information exchange program with the rest of the world community must also be very active in finding new ways and techniques to protect their vital information assets from unlawful use. The challenge before technologists, innovators, and decision makers is to reconcile the need to know and the need to protect the knowledge for competitive advantage.

2.5 ENVIRONMENTAL REQUIREMENTS FOR ENCOURAGING INDIVIDUAL INNOVATIVENESS

The environment, or organizational climate, can have impact on the innovativeness of individuals. A firm with programs that stimulate innovative activities is in a better position to provide products and services that will compete in the world market. In such a company, workers have access to broadly based information, have control over what they do, have freedom and encouragement to explore new ideas, and are provided tools for innovativeness. Here are 10 basic environmental requirements that facilitate the degree of innovativeness in an organization.

1. The organization personnel must understand the business environment in which they are operating.
2. There must be central overall goals and objectives toward which the total organization is striving. These goals and objectives must be understood by the whole organization.
3. The organizational goals must be broken down into subtasks and operations clearly understood by the whole organization.
4. The challenges and areas of new ventures for the organization must be communicated to the whole organization.
5. There must be freedom to generate and test new ideas within the scope of responsibility.
6. Cross-fertilization of ideas among departments, functional areas, and people must be encouraged.
7. A suggestion program should be instituted that allows those with new ideas to channel them for the benefit of the organization.
8. Opportunities must exist for individual growth and for individuals to implement new ideas for the organization.
9. There is appropriate reward for the contribution made by individuals through innovation of new products or services.
10. There is full employment and organizational stability for the workers.

An unstructured environment is usually good for the research and development component of the innovation process, which creates new ideas. But structured environments in manufacturing and marketing also permit process innovation. A fertile environment for innovators is one where (1) everyone within the organization feels capable of being a part of the decision-making process, (2) ideas and suggestions are welcome, (3) information sharing is encouraged, (4) participative style of management is present, and (5) there is timely reward for performance.

2.6 INDIVIDUAL INNOVATORS

The personal characteristics of the individual innovators are of primary importance in all phases of successful technological innovation. It is not easy to pinpoint the attributes and personality traits that guarantee a particular individual will be innovative. The basic educational training is neither a necessary nor a sufficient condition for identifying potential innovators. Those who did not go to school at all have on occasion come up with radical inventions. There is no sure explanation of the innovative process, but imaginative common sense blended with techno-scientific creativity and leadership traits is certainly necessary. Some individuals are most innovative when left alone, whereas others innovate effectively in a very interactive setting, but the following personal characteristics are usually present in successful innovators.

1. Good interpersonal skills in working with superiors, subordinates, peers, and other people in introducing technical concepts and solutions and in handling implementation of technical ideas.
2. Good vision and foresight, balanced judgment, self-confidence, and endurance.
3. Good understanding of the limitations of both natural phenomena and political settings. A basic understanding of the marketing requirements needed to take innovative ideas to fruition is also helpful.

2.7 THE IDEAL INNOVATIVE MANAGER

Likert (1967) isolated three variables that are representative of his total concept of participative management.

1. The use of supportive relationships by the manager.
2. The use of group decision making and group methods of supervision.
3. The manager's performance goals.

The manager's supportive relationship is characterized by the following:

1. Confidence and trust.
2. Interest in the subordinate's future.
3. Understanding of and desire to help with problems.
4. Training and helping the subordinate to perform better.
5. Teaching subordinates how to solve problems rather than giving the answer.

6. Giving support by making available the required physical resources.
7. Communicating information that subordinates must know to do their jobs and also the information they wish to know so that they may identify more with the operation.
8. Seeking out and attempting to use ideas and opinions.
9. Approachability.
10. Crediting and recognizing accomplishments.

The profile of the ideal innovative manager should include but not be limited to the following traits:

1. Has strong desire for innovative products or services. Wants people to come forward with new ideas and welcomes their implementations.
2. Has strong empathy when dealing with people, and possesses a caring attitude when dealing with the desires and needs of individuals and the organization.
3. Provides and encourages a trustworthy working environment where people can share ideas honestly.
4. Has a high level of creativity and is technically competent. Has a thorough knowledge of the business and has good ideas on how to improve it.
5. Is loyal and supportive of employee contributions and ideas, and works with them to obtain the resources needed for implementing such ideas.
6. Delegates work effectively and gives necessary control to workers to perform their tasks.
7. Accommodates failures, listens effectively, and rewards bad and good behavior in a timely manner.
8. Provides essential guidance when required; provides feedback on performance and monitors key activities effectively.
9. Is innovative and self-confident.
10. Is willing to take risks, pursue new ventures, and encourage subordinates to do the same.

Heinrich, Petersen, and Ross (1980) provide these guidelines for successful supervisory performance.

1. Deal with people as human beings, not machines.
2. Lead, do not drive or push.
3. Get people to like and respect you, create loyalty, win cooperation, instill confidence, build morale, and make people feel that they belong.
4. Listen to grievances.
5. Give credit when due, and time it psychologically.
6. Explain changes in advance.

7. Give orders clearly and precisely.
8. Ask for opinions and suggestions.
9. Be patient and impartial, consistent, friendly and courteous.
10. Display personal interest in the home life, hobbies, avocations, recreations, and personal problems of your workers.
11. Do not argue or be dogmatic when you disagree.
12. Get to know your own personal characteristics so as to avoid irritating or antagonizing others.
13. Get to know the personal characteristics, likes, dislikes, whimsies, convictions, idiosyncrasies, and motivating qualities and fundamental instincts of your workers.
14. Same as above for your boss, so as to enable you to get along with him.
15. Recognize your responsibilities to both management and labor.
16. Run your department as a business.
17. Find out what the workers really want most.
18. Test your subordinates to check attitude and ability.
19. Maintain a personal history record of each employee.
20. Put the "team" and competitive spirit to work.
21. Learn to recognize symptoms of trouble.
22. Correct misdemeanors only when a person has cooled off.
23. Anticipate difficulties and remove obstacles in advance; plan ahead and organize.
24. Interest the workers in the quality of production.
25. Keep in sound physical health and develop a saving sense of humor.

2.8 PREPARING FOR NEXT-GENERATION TECHNOLOGY

According to Edosomwan et al. (1988), companies have control over the current technology by which they design and produce products and services. However, technological change in large steps, commonly called *next-generation technology*, opens up competitive situations during the transition from current to next-generation technology. In that time, new companies are created, old companies fail, and competition expands or contracts market share and profitability. The problem is that no single company can create and acquire next-generation technology entirely by itself. Consequently, during such transitions companies must both depend on and be at the mercy of their competitors. Managing this transition is one of the important modern problems of technology management.

There is a growing consensus that the main purpose of corporate research is to broadly manage the technological future of the corporation. This idea is not really new—corporate research laboratories were originally established precisely to create the technological future of the corporations—but now the emphasis is on a broader vision of technology management, meaning (1) a better integration of business strategy, and (2) improving the organizational ability to act rapidly and effectively on technological change.

For business strategy, broadly managing the technological basis requires looking beyond simple financial practices for business diversification and planning. For manufacturing strategy, this requires a sophisticated focus on both product differentiation and low-cost manufacturing, while also creating shorter product development cycles. For competitive positioning, this requires planning the implementation of next-generation technology. Managing next-generation technology is one way to improve the effectiveness of corporate research.

Theorists have argued that organizations experience periods of relative stability interrupted by short strategic reorientations. Organizational change can be viewed as a kind of evolution, stimulated by and responding to change in business and economic environments. Effective top-level leadership for making strategic reorientation has long been recognized as important. Conversely, ineffective top-level leadership for making strategic reorientation is also important; ineffective top-level leadership is one of the major reasons for corporate decline across the technological discontinuities of next-generation technology.

Vigilance, vision, energy, and competence are the executive qualities necessary to accomplish strategic reorientations. A strategic team is required because managing next-generation technology requires several kinds of competence: technology, marketing, manufacturing, personnel, and finance. One major problem in strategic reorientation for next-generation technology is seeing through and creating consensus about all the organizational changes required. A second major problem is revising strategic reorientations as experience points out mistakes or weaknesses of the vision.

2.8.1 Impact of Poor Technology Strategy

There are several ways in which next-generation technology ignored by business strategy can put a company out of business.

1. The new technology provides products or services with superior functionality or performance to existing products.
2. The new technology provides existing products or services at a reduced price.
3. The new technology provides products or services that substitute for existing products or services with superior performance or reduced cost.

2.8.2 Recommended Approach to Preparing for Next-Generation Technology

1. Corporate research must strategically plan next-generation technology, and this is best done within an industrial/university/governmental research consortium.

40 REQUIREMENTS FOR INNOVATION AND TECHNOLOGY MANAGEMENT

2. Corporate research must tactically plan next-generation technology, and this is best done jointly with business divisions.
3. Marketing experiments must be set up and conducted jointly with corporate research and business divisions.
4. Computer-integrated manufacturing capability with flexibility is necessary for shortening the product development cycle and for cost-effective manufacturing of small volumes.
5. Long-term financial planning focused on technology and market strategy is required.
6. Personnel planning and personnel development are required to reorient knowledge bases and skill mixes into next-generation technology.
7. Organizational structure and culture should be designed to encourage innovation, foster entrepreneurial productivity, and accommodate strategic reorientation.

2.8.3 Requirements for Managing Current- and Next-Generation Technology

Organizations that want to begin managing technology must recognize that certain skills are necessary; if those skills are not available, they must be acquired. Professionals in both manufacturing and service environments should be given opportunities to broaden their horizons by gaining additional experience in a multiplicity of technology functions and by dealing with the users of their own technology. Such managers will instill the philosophy that technology functions exist for support of the business unit.

Managing technology requires technical managers with

1. An orientation that technology exists to serve the interests of the business and is pursued to make a contribution to the success of the business unit.
2. At least 10 years of solid technical contribution in research, development, and manufacturing while at the same time focusing on technology approaches that meet the needs of the marketplace.
3. A broad interest in and knowledge of science and engineering technologies; knowledge and understanding of where and how those technologies could enhance the performance of the business unit.
4. A fundamental grounding in management theory and practice as well as the related areas of finance, business law, quantitative methods, and so on.
5. A knowledge not only of the business unit but of the industry and its competitors—and total understanding of the global situation.
6. An innate desire for continuous learning and exploration; the ability to be a creative generalist who synthesizes business and technology.
7. The ability to work and communicate with people at all levels.

8. Integrity as the top priority for success.
9. Understanding of the three E's—effectiveness, efficiency, and economy.

Implementing the right approach to managing technology requires the support of top management, a clear understanding of the organization's technological strengths and weaknesses, and the actions and techniques required to sustain the organization and provide the right competitive goods and services.

SUMMARY

The degree of success achieved by a firm through technological innovation depends on the selection process for managers and other personnel, the technology policy and strategy selected by the firm, the resources available to the firm, the operating environment and marketing conditions, and the innovative climate that is in place through the operating policies, practices, and procedures. The innovator's immediate environment and those supervising his work can have impact on the degree of innovativeness. A healthy work environment that provides freedom for ideas and encourages their implementation can only breed more successful innovators and new ideas. A participative management approach that involves all levels of workers in decision making allows and encourages innovation.

Innovative managers are hard to find and not easy to hold on to. The innovative manager is one who truly cares about the total picture, is willing to take risks, and encourages subordinates to give their best to the organization. Such a manager takes great satisfaction in doing the right things for employees and the business. A new idea to an innovative manager means great opportunity for success.

QUESTIONS AND PROBLEMS

2-1 Discuss the senior management role in innovation and technology issues.

2-2 How would the six major areas of technology policy presented be implemented in an environment without participative management philosophy?

2-3 Develop your own framework for formulating a corporate technology policy.

2-4 Describe how the operating environment can affect an individual's degree of innovativeness.

2-5 What were the innovative characteristics behind the development of the steam engine, computer, and copy machine?

2-6 Assume you are the coordinator of a technology management program in a firm. How would you foster cooperation between industry and academia on R&D issues?

2-7 How would you describe the next generation of computer technology?

2-8 Present a step-by-step approach for implementing a new technology policy in a medium-size firm.

2-9 What are the innovation and technology management problems confronting U.S. companies?

2-10 Discuss the correlation between the amount of money invested on research and development and the profitability levels of a firm.

2-11 Describe an ideal innovative organization and identify its inherent characteristics.

2-12 How you go about promoting a new idea in an organization with 20,000 employees?

2-13 Assuming you are the plant manager of a medium-size firm that produces sport shoes to customer orders, what kind of policy would you implement within the organization to encourage innovation of new products?

2-14 How would you go about selecting an innovative manager from a group of potential job applicants, and what attributes will you use for the selection process?

2-15 What are the three variables required for total participative management?

2-16 How can an innovative manager improve the confidence and trust in his organization?

2-17 What are the problems involved in taking too much risk or not taking risk?

2-18 Identify and discuss some of the entrepreneurial qualities that are sources of potential liabilities in managing technology.

2-19 How should a manager or organization reward an innovative individual?

2-20 What types of training should be provided in an organization to improve the degree of innovativeness?

2-21 What programs should be in place to stimulate innovative ideas that spur new technologies at the enterprise level?

2-22 What are some of the difficulties involved in forming innovative/research teams between two emerging firms competing in the same market place?

3
COMPREHENSIVE DESIGN RULES FOR EMERGING TECHNOLOGY

This chapter presents and discusses concepts for optimizing the design of a new technology. The Technology-Aided Systems Design Optimization Cycle (TASDOC) is discussed. Ten design rules for emerging technology are provided as a guide for system designers. Finally, the outcome dimensions that are impacted when technologies are introduced and used in the work environment are presented.

3.1 TECHNOLOGY-AIDED SYSTEMS DESIGN OPTIMIZATION CYCLE

As shown in Figure 3.1, the total design optimization of technology systems can be achieved in four phass. The system information gathering phase: the system planning, assessment, and measurement phases; the system selection phases; and the system testing and improvement phase. Technology-Aided Systems Design Optimization Cycle (TASDOC) requires the system designer to have the end user in mind when designing the system. The designer must combine and use interdisciplinary knowledge to understand all issues involved in the interface between man and technology. Another important requirement is the understanding of the limitations of man and the various types of technology as well as the advantages of their interaction and the kinds of task they excel in. TASDOC can be viewed as a user-oriented framework for optimization of system design.

3.1.1 System Information-Gathering Phase

This phase involves the gathering of sufficient information about people, activities, operational requirements, environmental constraints and conditions, and total

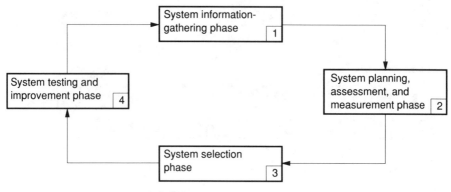

FIGURE 3.1 TASDOC phases.

TABLE 3.1 Positive Attributes of Humans and Machines

Humans	Machines
Strong ability to react to unplanned and unexpected low-probability event.	Strong ability to do different things at one time. Strong sensitivity to stimuli such as electronic waves.
Ability to detect signals in high noise levels.	Ability to respond quickly to control signals. Ability to perform deductive processes and their manipulation.
Strong ability to perform fine manipulation—for example, during unexpected misalignment.	Ability to repeat operations very rapidly, continuously, and precisely the same way over a long period.
Strong ability to continue to perform even when overloaded with work and stressful activities. Ability to perceive patterns and make generalizations about them.	Capability to operate in environments that are hostile to humans or beyond human tolerance.
Strong ability to reason inductively, to design logic in situations and systems, and to manipulate matters effectively.	Capability to perform complex and rapid computation with a high degree of accuracy.
Strong ability to exercise judgment when events cannot be completely defined.	Capability to store and recall large amounts of information in short time periods.
Strong sensitivity to a wide variety of stimuli; can detect certain forms of energy levels.	Capability to perform repetitive or very precise operation.
Ability to recall experience and use it wisely.	Insensitivity to extraneous factors.

material and resources needed for the system design. A thorough understanding of the types of activities in which humans and machines can excel is also very useful at this stage. Table 3.1 shows some of the abilities of humans and machines.

The information gathered at this phase enables the system designer to formulate an appropriate set of objectives for the impending design. Because information and facts are only as good as their sources, an attempt must be made to ensure that all design information is obtained from qualified personnel and the appropriate technical data base.

3.1.2 System Planning, Assessment, and Measurement Phase

At this phase, all alternative approaches for designing the system are explored. The approaches are examined to ensure that they meet the stated objectives. Factors such as product reliability, manufacturability, maintenance, economy, and labor force composition are weighed against their counterpart in each alternative approach. A standard of measurement is developed for each factor, including intangible factors. Tangible factors are quantified by values, indexes, or monetary terms.

3.1.3 System Selection Phase

The system designer selects the design that has the greatest potential to optimize all the factors identified in the first two phases. The system is designed with strong consideration of all aspects of design rules and principles (discussed in more detail in Section 3.2). To optimize system design, aspects of human engineering such as vision control, workplace layout, and motion analysis must be considered. Woodson and Conover (1964), Ayoub (1966), Kunz (1971), Khalil (1976), and McCormick (1976) discuss useful principles for engineering and system design requirements. The selected design should be the one that optimizes human safety at the workplace, productivity, quality, comfort, job satisfaction, and the quality of working life.

3.1.4 System Testing and Improvement Phase

At this phase, models of the designed system are tested and evaluated against the originally stated objectives. It is recommended that the final design be tested in an actual mode of operation. This is usually accomplished through a pilot project. Although testing a design through a pilot project may be expensive, it helps prevent a high failure rate that might be harmful in the long run. Improvements should be made, based on the pilot project test results. To ensure that all aspects of the requirements for the design are covered, it is recommended that a comprehensive design checklist be used; the computer-aided manufacturing ergonomic checklist (CAMEC) is presented in Appendix A.

TASDOC required ongoing follow-up actions through customer surveys or user inputs that feed back information on system performance. The TASDOC process should be viewed and implemented as a continuous one that interrelates system design actions in one model cycle to the design of a new system model. The key to successful implementation is teamwork among all functional areas of the organization. One approach that has worked well in recent years is the concept of early manufacturing involvement (EMI) in the design process. EMI requires designers, manufacturers, and potential users of technology to be involved in all phases of the design cycle. The communication between all functions allows for potential design problems to be corrected on a timely basis.

3.2 DESIGN RULES FOR APPROPRIATE TECHNOLOGY

A careful review of the OSHA report on man–machine related accidents indicates that for 620 accidents in six major firms between 1982 and 1986, about 79.6% of the accidents were due to design problems and 21.4% were caused by operator-induced errors. Even if machine-related accidents are minor, they point to a need for appropriate adherence to design rules and principles. In order to design technology appropriate for human comfort, reduced social and psychological costs, reduced work hazards, and increased productivity, the following design rules for emerging technology are provided as a guide for system designers.

3.2.1 Design Rule 1: People are the Center of Design

Systems designers are often caught in the trap of designing for the "average person." The average person, however, exists only as a statistical tool. The population of users of the various types of technology covers a very wide spectrum of several different body dimensions. As shown in Figure 3.2, the human skeletal system comprises components that have limitations in movement, endurance, and flexibility. It is therefore important for system designers to recognize human anatomical structure and to obtain accurate anthropometric dimensions to fit tasks to people. Table 3.2 shows selected structural body dimensions and weights of adults that must be considered in system design. Daniels et al. (1953), Daniels (1958), Woodson and Conover (1964), and Woodson (1981) have done detailed analyses of dimensional elements that must be considered for females. The selected dimensions of the female human body (ages 18 to 45) useful for initial design of workplace and equipment are presented in Figure 3.3 and Table 3.3. Similar dimensions for the male human body (ages 18 to 45) are presented in Figure 3.4 and Table 3.4. Failure to recognize human capacities, capabilities, and limitations in the design stage can result in psychological stress and fatigue for workers.

3.2.2 Design Rule 2: Use Principles of Kinesiology in the Design

Any design that will result in incompatible movements should be avoided. Designers should allow free movement of joints around joint axes for system-oriented controls. Khalil (1976) has illustrated the successful application of this

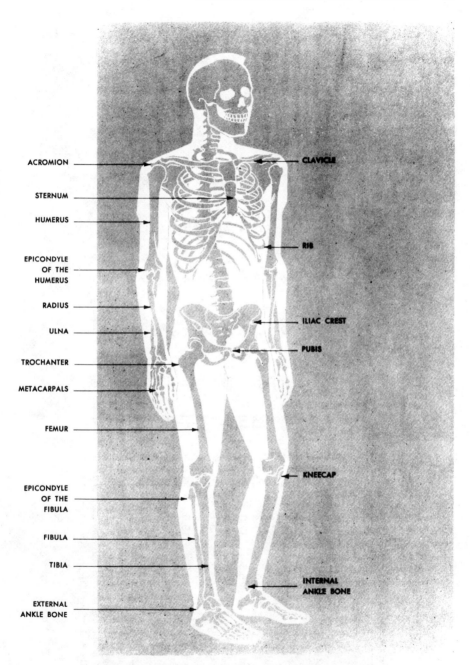

FIGURE 3.2 Components of the human skeletal system

TABLE 3.2 Selected Structural Body Dimensions and Weights of Adults

	Dimensions: in						Dimensions: cm[b]					
	Male, percentile			Female, percentile			Male, percentile			Female, percentile		
Body feature	5th	50th	95th	5th	50th	95th	5th	50th	95th	5th	50th	95th
1 Height	63.6	68.3	72.8	59.0	62.9	67.1	162	173	185	150	160	170
2 Sitting height, erect	33.2	35.7	38.0	30.9	33.4	35.7	84	91	97	79	85	91
3 Sitting height, normal	31.6	34.1	36.6	29.6	32.3	34.7	80	87	93	75	82	88
4 Knee height	19.3	21.4	23.4	17.9	19.6	21.5	49	54	59	46	50	55
5 Popliteal height	15.5	17.3	19.3	14.0	15.7	17.5	39	44	49	36	40	45
6 Elbow-rest height	7.4	9.5	11.6	7.1	9.2	11.0	19	24	30	18	23	28
7 Thigh-clearance height	4.3	5.7	6.9	4.1	5.4	6.9	11	15	18	10	14	18
8 Buttock-knee length	21.3	23.3	25.2	20.4	22.4	24.6	54	59	64	52	57	63
9 Buttock-popliteal length	17.3	19.5	21.6	17.0	18.9	21.0	44	50	55	43	48	53
10 Elbow-to-elbow breadth	13.7	16.5	19.9	12.3	15.1	19.3	35	42	51	31	38	49
11 Seat breadth	12.2	14.0	15.9	12.3	14.3	17.1	31	36	40	31	36	43
12 Weight[a]	120	166	217	104	137	199	58	75	98	47	62	90

[a]Weight given in pounds (first six columns) and kilograms (last six columns).
[b]Centimeter values rounded to whole numbers.
Source: From *Weight, height, and selected body dimensions of adults: 1960–1962.* Data from National Health Survey. USPHS Publication 1000, series 11, no. 8, June, 1965.

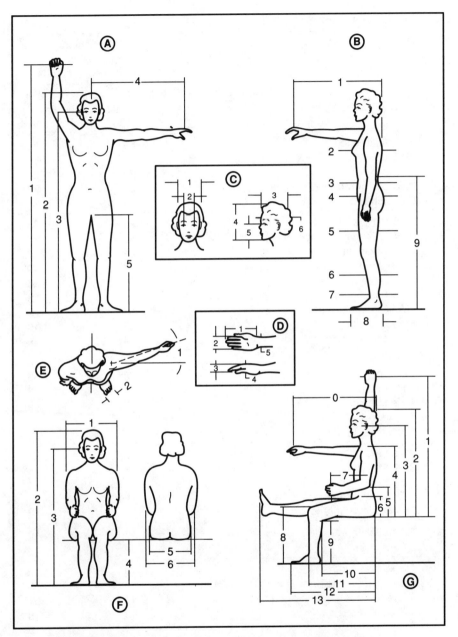

FIGURE 3.3 Illustrations of female human body dimensions.

TABLE 3.3 Female Human Body Dimensions

	Dimensional Element Weight	5th Percentile 102 lb	95th Percentile 150 lb
A	1 Vertical reach	69	81
	2 Stature	60	69
	3 Eye to floor	56	64
	4 Side arm reach from centerline of body	27	38
	5 Crotch to floor	24	30
B	1 Forward arm reach	24	35
	2 Chest circumference (bust)	30	37
	3 Waist circumference	24	29
	4 Hip circumference	33	40
	5 Thigh circumference	19	24
	6 Calf circumference	12	15
	7 Ankle circumference	8	9
	8 Foot length	9	10
	9 Elbow to floor	34	46
C	1 Head width	5	6
	2 Interpupillary distance	2	3
	3 Head length	6	7
	4 Head height	—	9
	5 Chin to eye	—	4
	6 Head circumference	20	23
D	1 Hand length	6	7
	2 Hand width	3	4
	3 Hand thickness	1	1
	4 Fist circumference	9	11
	5 Wrist circumference	6	7
E	1 Arm swing, aft	40°	40°
	2 Foot width	3	3
F	1 Shoulder width	13	19
	2 Sitting height to floor (standard chair)	45	55
	3 Eye to floor (standard chair)	41	51
	4 Standard chair	18	18
	5 Hip breadth	13	15
	6 Width between elbows	11	23
	0 Arm reach (finger grasp)	22	33
	1 Vertical reach	39	50
	2 Head to seat	27	38
	3 Eye to seat	25	32
	4 Shoulder to seat	18	25

TABLE 3.3 (Continued)

	5 Elbow rest	4	12
G	6 Thigh clearance	4	6
	7 Forearm length	14	18
	8 Knee clearance to floor	17	22
	9 Lower leg height	14	19
	10 Seat length	13	23
	11 Buttock–knee length	18	27
	12 Buttock-toe clearance	27	37
	13 Buttock–foot length	34	49

[a]All except critical dimensions have been rounded to the nearest inch.

rule in technology and workplace design. Failure to recognize this rule could cause stress on the human skeletal system during operation of machine controls.

3.2.3 Design Rule 3: Use Physiological Criteria for Design

Figure 3.5 illustrates some problems that incompatible design can cause. Limitations imposed by the input-output channels of human-operated machines restrict an operator's interaction with his environment. The workplace envelope, input-output channels, control sensors, illumination levels, signs, and seating requirements must be compatible with the anthropometric dimensions of the individuals in the workforce who will be using the proposed technology, tool, or equipment. Woodson and Conover (1964) recommend that dimensions of the large worker should be used for determining clearances, whereas dimensions of the small worker should be used to determine limits of reach. They further suggest that all layouts be verified in a three-dimensional mock-up where live subjects representing the extremes of the expected population can try out the layout. Stress concentration points in mechanical design can be minimized by recognition of the dynamic changes brought about by normal bending, slumping, moving, lifting, and stretching. Other factors such as clothing, weather, and environmental humidity should also be considered in designing the human-machine interface.

3.2.4 Design Rule 4: Apply Psychological Principles to Improve Morale and Increase Job Satisfaction

New technology is often introduced into the work environment without an analysis of its impact on the morale of workers and the quality of working life. The use of psychological tests to analyze problems is recommended whenever appropriate. The job description index, a rating scheme, is also recommended for evaluating the impact on new technology on workers' and supervisors' satisfaction, pay, and promotion. Edosomwan (1985a, 1985b, 1986b) demonstrates the use of the comparative judgment instrument (CJI) to measure job satisfaction and psychological stress in situations where new technologies such as computers and robotic devices

TABLE 3.4 Male Human Body Dimensions

	Dimensional Element Weight	Dimension (in inches except where noted)[a]	
		5th Percentile 132 lb	95th Percentile 201 lb
A	1 Vertical reach	77	89
	2 Stature	65	73
	3 Eye to floor	61	69
	4 Side arm reach from centerline of body	29	34
	5 Crotch to floor	30	36
B	1 Forward arm reach	28	33
	2 Chest circumference (bust)	35	43
	3 Waist circumference	28	38
	4 Hip circumference	34	42
	5 Thigh circumference	20	25
	6 Calf circumference	13	16
	7 Ankle circumference	8	10
	8 Foot length	10	11
	9 Elbow to floor	41	46
C	1 Head width	6	6
	2 Interpupillary distance	2	3
	3 Head length	7	8
	4 Head height	—	10
	5 Chin to eye	—	5
	6 Head circumference	22	24
D	1 Hand length	7	8
	2 Hand width	4	4
	3 Hand thickness	1	1
	4 Fist circumference	11	12
	5 Wrist circumference	6	8
E	1 Arm swing, aft	40°	40°
	2 Foot width	4	4
F	1 Shoulder width	17	19
	2 Sitting height to floor (standard chair)	52	56
	3 Eye to floor (standard chair)	47	52
	4 Standard chair	18	18
	5 Hip breadth	13	15
	6 Width between elbows	15	20
	0 Arm reach (finger grasp)	30	35
	1 Vertical reach	45	53
	2 Head to seat	34	38
	3 Eye to seat	29	34
	4 Shoulder to seat	21	25

TABLE 3.4 (Continued)

	5 Elbow rest	7	11
G	6 Thigh clearance	5	7
	7 Forearm length	14	16
	8 Knee clearance to floor	20	23
	9 Lower leg height	16	18
	10 Seat length	15	22
	11 Buttock–knee length	22	37
	12 Buttock-toe clearance	32	37
	13 Buttock–foot length	39	46

[a]All except critical dimensions have been rounded to the nearest inch.

were introduced into the manufacturing environment. Here are some of the questions in the CJI instrument.

Question: Which task is more stressful?
( ) Task R (Robot) ( ) Task M (Manual) ( ) Task C (Computer)
Question: Which task requires working with the greatest speed?
( ) Task R (Robot) ( ) Task M (Manual) ( ) Task C (Computer)
Question: Which task provides more decision latitude?
( ) Task R (Robot) ( ) Task M (Manual) ( ) Task C (Computer)

In order to show how much one task is preferred to the other, write M, R, and C in the appropriate space below:

(1)	(2)	(3)	(4)	(5)
no impact	little impact	moderate impact	fairly high impact	high impact

More CJI questions can be found in Appendix B. It is recommended that the CJI instrument be administered before and after the introduction of a new technology in the workplace. The scale used in the CJI lends itself to statistical analysis to provide confidence levels for results.

3.2.5 Design Rule 5: Recognize the Worker's Right over Control of Work Activity

In both repetitive and nonrepetitive tasks, designers often make the mistake of designing mechanical controls or sensors to control the worker's movement, freedom, and pace of work. The study conducted in a computer-aided manufacturing environment by Edosomwan (1985a, b, c) shows that when the pace of the production worker was controlled by a fixed computer speed, this led to decrease in job satisfaction and psychological stress for the workers. The same case study also

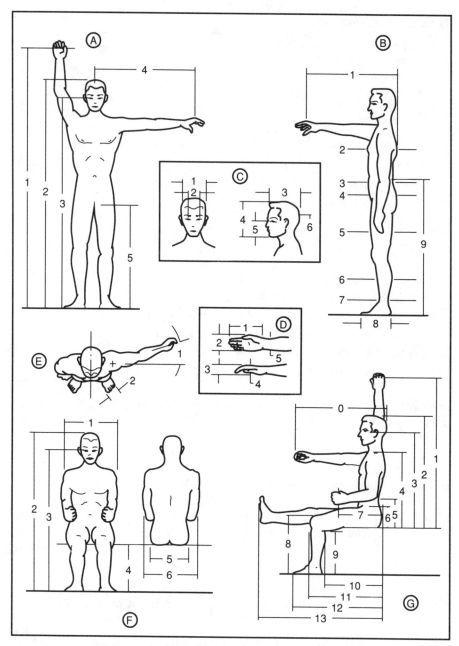

FIGURE 3.4 Illustrations of male human body dimensions

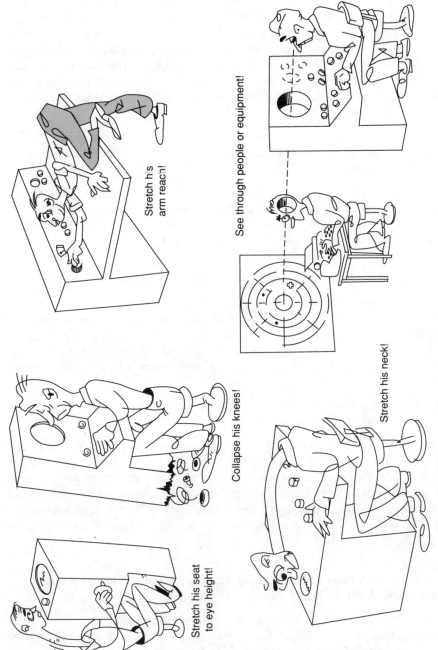

FIGURE 3.5 Results of some typical design errors on the human operator.

found that in situations where computer variable speed is allowed, job satisfaction improves and psychological stress decreases. Most technology developed without sufficient planning takes control away from workers in order to minimize production errors. The tasks performed with such technology become routine and repetitious, leading to boredom and stress. Technology applications in the workplace should be designed and implemented to allow workers to have decision latitude; otherwise, they may become counterproductive to the goals for which they are developed. A case study that illustrates the importance of ergonomic factors in design is presented in Chapter 8.

3.2.6 Design Rule 6: Allow for Skill Development through the Use of New Technology

Technology applications should allow workers the opportunity to develop new skills in their work. Often, technology applications create dead-end jobs. The best application of a technology allows variety in the various task performed, learning experience, and expansion capabilities for different applications.

3.2.7 Design Rule 7: Design User-Friendly Systems

Technology such as computers and expert systems should be designed to be user-friendly. The simpler the programming code in a particular computer application, the less irritation and hassle a user has to put up with. The response time of systems, codes, commands, screen checks, push buttons, help screens, and other system parameters must be designed to allow the maximum human-friendly interaction.

3.2.8 Design Rule 8: Involve the Users of Technology-based Systems during the Design Phase

The total team involvement of system designers, manufacturers, and users in system design issues enables potential problems to be corrected in the early part of the design cycle. This involvement helps system designers become aware of the human factor implications early in the game and provides consensus on desirable goals. The more compatible technology-based systems are with the job satisfaction of the workers and structure of the production environment, the more likely they are to be successfully adopted.

3.2.9 Design Rule 9: Give the Technology-based Systems Ability to Advise, Alert, or Warn Users of Potential Events

Three case studies in computer-aided manufacturing environment and robotic device applications by Edosomwan (1985a, b, c, ; 1986c, d) have shown that providing control sensors in computer-aided devices and error-recovery sensors in robotic devices reduces the potential for accidents by 87%. This further implies that designers of technology-based systems must find a compromise between those aspects of the system that influence system productivity and production quality, and those that affect the quality and safety of the task performed by production workers. Human comfort and safety should not be sacrificed for productivity gains through the use of technology.

3.2.10 Design Rule 10: Implement a Total Productivity and Quality Improvement System

Unless ongoing productivity and quality improvement efforts are applied to technology designs, the users of the systems may not be able to afford them. To reduce the cost of the technology to users, it is necessary to:

- Identify the resources needed for technology development.
- Select preliminary technical concepts in design.
- Apply conventional concepts in design.
- Evaluate requirements and opportunities for new or improved concepts in design.
- Record cost estimates and measure total, total factor, and partial productivities.
- Use of value engineering extensively to simplify operations and minimize subsystem components that are required for the technology development.

To improve the quality of technology for users it is necessary to:

- Implement improvement techniques that minimize the number of engineering changes in technology-based systems.
- Implement statistical process control in the design and manufacturing phases of a technology life cycle.
- Place emphasis on achieving quality at the source of design and production through the use of quality improvement teams, quality error-removal techniques, and ongoing process monitoring and improvement.
- Evaluate the reliability of technology-based systems at the design, manufacturing, and user stages.
- Assess the system's potential impact on environment, people, and the quality of working life.

At both the R&D and design engineering phases, a careful analysis of cost/performance-effectiveness should be made. A total productivity assessment should be done at all phases of the technology life cycle. In Chapter 4 a technology-oriented total productivity measurement system that can be used for this purpose is presented along with other cost/performance ratios. In the construction, testing, and installing phase of technology development, attention must be focused on design details; proper calibration and repeated tests must be performed on all components,

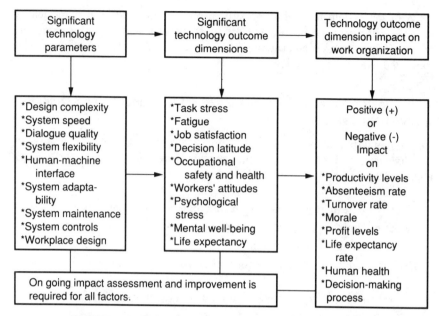

FIGURE 3.6 Technology quality of working life model (TWLM).

subassemblies, and completed systems or equipment. The implementation of a total productivity and quality improvement effort at all phases of the technology life cycle requires upper management support and teamwork among all members of the organization.

3.3 UNDERSTANDING TECHNOLOGY'S IMPACT ON THE QUALITY OF WORKING LIFE

Studies conducted by Kling (1978), Danzinger and Dutton (1977), Turner (1980), Bradley (1977), Karasek and Turner (1981), and Edosomwan (1985a, b, c and 1986b) have shown that technology design parameters, associated with task environment and other production or service variables have a significant impact on the quality of working life. In order to improve productivity in the work environment, the technology working life model (TWLM) outcome dimensions shown in Figure 3.6 must be managed continuously through all phases of the technology life cycle. TWLM suggests that technology parameters such as design complexity, task complexity, speed, dialogue quality, and system flexibility are likely to affect human beings positively or negatively in the work environment. Changes in workers' attitudes, psychological stress, job satisfaction, fatigue level, and decision latitude and control are some of the outcome dimensions that can easily be affected by the introduction and use of any type of technology.

Changes in technology are also bound to have positive or negative impact on the work organization. Productivity levels, absenteeism rate, turnover rate, morale, profit levels, and life expectancy rates are some of the many variables likely to be affected by technology outcome dimensions.

An ongoing teamwork approach between the designer and users of technology is needed to improve and resolve the potential negative impact of technology outcome dimensions at all phases of the technology life cycle. The need to understand and manage the impact of technology on the quality of working life can best be explained by the result of case studies presented in Chapter 8.

SUMMARY

This chapter has discussed a formalized procedure for achieving optimal design of technology-based systems. The 10 design rules have been presented as a guide to system designers, manufacturers, and users. There is no doubt that the greatest asset for any organization is its human resources; without human resources the development and use of technology will not happen in the first place. It is therefore important that the framework presented in this chapter be used to design and implement technology-based systems to improve productivity and the quality of working life.

QUESTIONS AND PROBLEMS

3-1 What are the problems involved in implementing the technology design optimization cycle?

3-2 How can a systems designer optimize the user-friendliness of new technology-based systems?

3-3 What strategy would you adopt to involve the users of technology in the design phase?

3-4 What are the potential implications of not following the comprehensive design rules for emerging technologies?

3-5 How should the total cost of technology be reduced for competitiveness?

3-6 What are the limitations of the human skeletal system in using technology-based systems?

3-7 What are the problems involved in designing technology-based systems for the average human?

3-8 How would you access the impact of technology-based systems on the quality of working life and job satisfaction?

3-9 How would you design a technology-based system to allow for human flexibility and skill development?

3-10 What measurement and evaluation techniques are feasible for assessing the effectiveness of new technology-based systems?

3-11 How would you go about managing the apparently contradictory sets of objectives related to technology outcome dimensions such as productivity, quality, job satisfaction, stress, safety, and the quality of working life?

3-12 Assuming you are the systems manager in charge of designing commercial aircraft, how would you go about implementing the comprehensive design rule to optimize the aircraft design and ensure consumer safety and overall satisfaction?

3-13 Discuss the problems associated with using subjective indexes in evaluating the potential impact of a new technology.

3-14 How should trade-off analysis be handled when performing technology assessment and evaluation? What procedures should be used in handling intangible factors involved in technology utilization?

3-15 Discuss some of the techniques that should be used to reduce the total cost of technology to users.

4
JUSTIFYING AND IMPLEMENTING TECHNOLOGY

Technology justification is not only a matter of looking at cost components and payback period. The justification process for a new technology requires that both tangible and intangible factors be addressed before a given technology is selected for implementation in the workplace. Most technology or equipment selection is done using partial measures such as labor productivity and return on investment. This chapter presents quantitative and qualitative techniques for justifying a new technology as part of a technology-oriented total productivity measurement model that takes into account all the measurable inputs and outputs. A weighted scoring model that takes intangible factors into account is offered. Quantitative measures for assessing the level of technological progress are discussed. This chapter also presents a step-by-step approach for implementing a new technology. The 6C principles for managing technology projects and the role of the project manager in managing technology are described. The common implementation problems are discussed, as is overall strategy for overcoming them.

4.1 UNDERSTANDING THE UNCERTAINTIES IN CAPITAL DECISIONS

There are various uncertainties involved in capital decisions, especially those related to the acquisition of new technology or equipment. The technology justification process offers a comfort in weighting the benefits against the cost and risk involved in acquiring a new technology. It is important for decision makers to be aware that no matter what type of tool or equipment is finally selected after justification, the following uncertainties will always be present in a competitive world economy.

1. Size, share, and growth rate in the market. These can be affected by several factors, including new competitors, effectiveness of marketing efforts and techniques, growth of economy, new products, aggregate consumer demand, product quality, and alternative products.
2. Operating and fixed costs for producing goods and services. Factors such as overhead rates, administrative expenses, learning curve, labor rates, energy expenses, and process improvements can influence the amount of risk involved in managing fixed and operating costs.
3. Intangible factors such as customer, employee, and public reactions.
4. Obsolescence and deterioration, which can affect the useful life of tools and technologies.
5. Major technical breakthroughs such as process technology or the ability to substitute materials and products, which can affect product development risk.
6. Factors such as price changes, competition level, time delay of prices, quality of estimates, machine loadings, capacity statements, and quantity of production, all of which may influence selling price and required investment risk.
7. Demand rate for scrap and used tools and machinery, which affects residual value.

4.2 A PERSPECTIVE ON THE TECHNOLOGY JUSTIFICATION PROBLEM

Hunter (1985) has reviewed the literature on the justification dilemma involved in flexible manufacturing systems. The common themes extracted are as follows:

1. The first prerequisite of successfully justifying a flexible automation project is the development of a top-down, long-range automation strategy.
2. Financial evaluation should be geared to evaluation of the total "system."
3. A detailed understanding of current manufacturing costs, particularly overhead, should be a prerequisite for a flexible automation proposal.
4. Current accounting methods do not provide the necessary information to readily understand the true manufacturing cost.
5. Developing and selling the proposal for a flexible system is a nontrivial assignment that requires an individual with a broad understanding of manufacturing, particularly engineering and accounting.
6. Multi-disciplinary teams are needed to develop and evaluate flexible manufacturing systems.
7. Economic evaluation of a flexible manufacturing system should include the best estimates available for all benefits, tangible and intangible. Failure to do so implicitly assigns a value of zero to that benefit.

Hunter further points out that justification of appropriate flexible manufacturing systems will be made easier by:

1. Top-down initiation.
2. An understanding of current overhead costs before the system is designed.
3. Project manager with a broad manufacturing background.
4. System designed by a multidisciplined team.
5. Early proactive involvement by finance.

Kaplan (1986) points out that one of the difficulties encountered in the justification of new technology is the use of unrealistically high discount rates to cushion the firm or analyst from the risk associated with high-technology innovations.

It is quite difficult to find reusable methods for justifying technology in both manufacturing and service industries. This notion is supported by Simpson (1984) in a study of high-technology (robotic) investment justification. The degree of formality, originality, and computerization of economic analysis varied widely.

In order to correct the deficiencies in the traditional methods of justifying new technologies, several approaches have been recommended in recent years. These approaches vary from the inclusion of nonmonetary intangible factors to the consideration of the impact of technology on market share, product quality, quality of working life, and morale improvement. The next several sections of this chapter discuss both the qualitative and quantitative techniques recommended for justifying technology.

4.3 THE PAYBACK PERIOD JUSTIFICATION TECHNIQUE

The payback period (PBP) justification technique is based on the comparison of the payback period of a new technology with the expected lifetime of the technology or equipment. A simple payback period for a technology is defined as the investment cost divided by the net annual savings. The generally accepted rule of thumb is:

A payback period of less than one-half the lifetime of a technology or equipment would generally be considered a viable investment where the lifetime is ten years or less. Otherwise reject new technology.

Definitions

PBP_i = Payback period for a typical technology or equipment i.

IC_i = Initial cost of technology or equipment i.

B_i = Annual benefits involved in using technology or equipment i.

C_i = Annual cost involved in using technology or equipment i.

$CF_i = B_i - C_i$ = Annual cash flow involved in using technology or equipment i.

64 JUSTIFYING AND IMPLEMENTING TECHNOLOGY

Case One. If equal annual cash flows can be assumed, then

$$\text{PBP}_i = \frac{\text{IC}_i}{\text{CF}_i} \tag{4.1}$$

Case Two. For unequal annual cash flows, add each annual cash flow beginning with year 1 until the sum reaches IC_i. The number of years added will be the payback period.

This technique has the major disadvantage of failing to consider annual cash flows beyond the payback period. In situations that require the benefits of the implementation of the new technology or equipment to be analyzed beyond the payback period, this technique provides no useful assistance. The simple payback period criterion for economic analysis fails to account for the time value of money and ignores income or savings that occur after the payback period. Although this technique is widely used for tactical and operational decisions, it is not recommended for technology justification because of these disadvantages.

4.4 THE RETURN ON INVESTMENT (ROI) JUSTIFICATION TECHNIQUE

This technique takes into account the depreciation of the technology or equipment over its useful economic life. In addition, the average annual benefits and costs of the new technology or equipment are considered. ROI is usually expressed in percent per year. The rate of return is defined as the percentage or rate of interest earned on the unrecovered balance of an investment. The mathematical expressions for ROI computation are presented below: Letting:

IC_i = Initial cost of technology or equipment i.

DC_i = Depreciation cost per year of technology or equipment i.

AAB_i = Average annual benefits of technology or equipment i.

AB_i = Annual benefits of technology or equipment i.

N_i = Useful economic life of technology or equipment i.

ROI_i = Return on investment from technology or equipment i.

We obtain the following relationships:

$$\text{DC}_i = \frac{\text{IC}_i}{N_i} \tag{4.2}$$

$$\text{AAB}_i = \frac{\sum_{i=1}^{N_i} AB_i}{N_i} \tag{4.3}$$

$$\text{ROI}_i = \frac{\text{AAB}_i - \text{DC}_i}{\text{IC}_i} \qquad (4.4)$$

Most firms set their own minimum return on investment. Depending on the industry and types of technology, acceptable ranges of ROI can vary from 15–40%. Although the ROI technique is better than payback period technique, it also has a disadvantage of not considering cash flows in the computational procedure. This restricts its use in evaluating long-term, large-scale investments or in evaluating technology and equipment with unusually long economic life.

4.5 NET PRESENT VALUE (NPV) JUSTIFICATION TECHNIQUE

This technique is also known as the present worth method. The procedure for obtaining the net present value (NPV) of a typical technology or equipment is as follows:

1. Compute or find in a present value table the present value of the cash flows generated by a given investment in a particular technology.
2. Sum the discounted cash flows for each year and subtract the initial cost to obtain the NPV for the technology or equipment concerned.

TABLE 4.1 Net Present Value Example

Consider a case in which capital investment for a new computer technology will cost $50,000, and the annual savings obtained from the use of computer are expected to decrease from $22,500 to $5,000 as shown over the four-year useful economic life of the computer. The XYZ company's minimum rate of return on investment is 10%.

Year	Benefits	Costs	Cash Flow[a]	PV[b] of $1 @10%	Discounted Cash Flow
0	$0	$(50,000)	$(50,000)	1.000	$(50,000)
0–1	22,500	0	22,500	0.952	21,400
1–2	22,500	0	22,500	0.861	19,400
2–3	15,000	0	15,000	0.779	11,700
3–4	5,000	0	5,000	0.705	3,500
Total	$65,000	$(50,000)	$15,000		$6,000 NPV

The proposed purchase recovers all costs, the 10% opportunity costs of the firm's capital is realized, and the investment will yield an additional $6,000 return. Therefore the proposal should be accepted.

[a] Cash flow for each year is obtained by Annual cash flow = Benefits − Costs.
[b] $\text{PV} = (1+i)^{-t}$ is obtained from interest tables.

JUSTIFYING AND IMPLEMENTING TECHNOLOGY

The mathematical expression for calculating the net present value of all of the expected cash flows by discounting them by required rate of return is as follows. Let:

F_t = Future cash flow in period t
P = Present or initial investment
i = The required rate of return
NPV = Net present value

Then

$$\text{NPV} = -P + \sum_{t=1}^{n} F_t(1+i)^{-t} \qquad (4.5)$$

See Table 4.1 for an example using this procedure. When evaluating individual technology projects, accept a project if the present worth is positive. When mutually exclusive technology projects are involved, the criterion for the selection should be the largest present worth. It is important to point out that present worth comparisons of technology projects are feasible only for coterminated projects with equal life.

4.6 INTERNAL RATE OF RETURN (IRR) JUSTIFICATION TECHNIQUE

This technique provides a basis for explicit comparison of the technology or equipment project's return with the discount rate appropriate for the company in justifying the use of such technology or equipment. The IRR technique complements the NPV technique. However, the IRR technique requires the use of a trial-and-error iterative process converging on a solution. The IRR is defined as the discount rate

TABLE 4.2 Internal Rate of Return Example[a]

Year	Cash Flow	PV[b] of $1 @ 10%	Discount Cash Flow	PV[b] of $1 @ 18%[c]	Discounted Cash Flow
0	$(50,000)	1.000	$(50,000)	1.000	$(50,000)
0–1	22,500	0.952	21,400	0.915	20,600
1–2	22,500	0.861	19,400	0.764	17,200
2–3	15,000	0.779	11,700	0.639	9,600
3–4	5,000	0.705	3,500	0.533	2,600
Total	$ 5,000		$ 6,000 NPV		$ 0 NPV

[a]Initial investment, annual cash flows, and minimum annual rate of return are the same as in Table 4.1.
[b]PV obtained from interest tables.
[c]The IRR, obtained by interpolation, is 18%, which is greater than the 10% minimum rate of return; therefore, the company invests on the view computer.

that reduces the stream of cash flows associated with the technology or equipment to a present value of zero. Table 4.2 presents an example of the IRR technique.

4.7 TOTAL PRODUCTIVITY JUSTIFICATION TECHNIQUE

The traditional approach of using labor productivity (output per labor-hour) and the rate of return on investment (ROI) to justify technology applications is not sufficient to help decision makers in understanding the total cost of doing business with a new technology or equipment. A technology-oriented total productivity measurement model (TOTPMM) developed by Edosomwan (1987a,b,c) is presented here as a more useful approach for justifying a new technology or equipment. Here are some important terms that will be used in the presentation:

Productivity Measures. The productivity measures provide the basis to compare total output obtained, or expected to be obtained, from a particular technology or equipment to the total or partial input utilized or expected to be utilized.

Total Productivity. This is the ratio of all measurable output to all measurable input.

Total Factor Productivity. This is the ratio of all measurable output to the sum of associated labor and capital input factors.

Partial Productivity. This is the ratio of all measurable output to one class of input. The input and output components of the TOTPMM are presented in Figures 4.1 and 4.2, respectively.

Notations for Derivation of Productivity Values and Indexes

i = Technology type ($i = 1, 2, 3, \ldots, m$), where m = number of types.
j = Technology life-cycle phase ($j = 1, 2, 3, 4$).

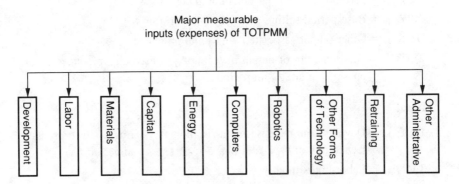

FIGURE 4.1 Input components considered in the technology-oriented total productivity measurement model (TOTPMM).

68 JUSTIFYING AND IMPLEMENTING TECHNOLOGY

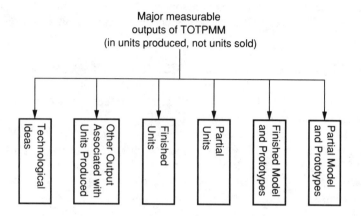

FIGURE 4.2 Output components considered in the technology-oriented total productivity measurement model (TOTPMM).

k = Technology development/manufacturing/service site ($k = 1,2,3,...,n$), where n = number of sites.

t = Study period ($t = 1, 2, 3, ..., p$), where p = number of periods.

Output

$\text{TCI}_{ij\,kt}$ = Total value of technological ideas of type i, produced in phase j, at site k, in period t.

$\text{FMP}_{ij\,kt}$ = Total value of finished model and prototype of technology type i, produced in phase j, at site k, in period t.

$\text{PMP}_{ij\,kt}$ = Total value of partially completed technology model and prototype.

$\text{FUP}_{ij\,kt}$ = Total quantity of finished units of technology.

$\text{PUP}_{ij\,kt}$ = Total quantity of partial units of technology.

$\text{PCT}_{ij\,kt}$ = Percentage of partial units completed.

$\text{SPT}_{ij\,kt}$ = Base period selling price per unit for a unit.

$\text{OOT}_{ij\,kt}$ = Other output associated with units produced.

$\text{TOT}_{ij\,kt}$ = Total quantity of output of technology of type i, produced in phase j, at site k, in period t.

Input

$\text{DET}_{ij\,kt}$ = Development expense input (in monetary terms) used to produce technology of type i, in phase j, at site k, in period t.

$\text{LRT}_{ij\,kt}$ = Labor-hours input used.

$\text{LRR}_{ij\,kt}$ = Labor rate per hour used.

$\text{MET}_{ij\,kt}$ = Materials expense.

CET_{ijkt} = Capital-related expense (includes fixed and working capital such as cash, accounts receivable, tools, plant, buildings, etc.). Capital is computed using lease-value concept.

EET_{ijkt} = Energy-related expenses (includes electricity, solar energy, water, coal, gas, etc.).

CRE_{ijkt} = Variable computer-related expense.

RRE_{ijkt} = Variable robotics-related expense.

OFT_{ijkt} = Variable expense from other forms of technology.

RET_{ijkt} = Retraining expense.

OET_{ijkt} = Other administrative expense.

TIT_{ijkt} = Total input used to produce technology to type i, in phase j, at site k, in period t.

Other Variables:

TBT = Base period time, which is the reference period to which output and input in monetary terms are reduced. Thus, the total productivity of a task expressed as dollar output/dollar input is in constant dollars.

TPT_{ijkt} = Total productivity of technology of type i, in phase j, at site k, in period t.

TFT_{ijkt} = Total factor productivity of technology of type i, in phase j, in site k, in period t.

PPT_{ijkt} = Partial productivity of technology of type i, in phase j, in site k, in period t.

$$TOT_{ijkt} = FMP_{ijkt} + TCI_{ijkt} + PMP_{ijkt} + (FUP_{ijkt})(SPT_{ijkt}) \\ + (PUP_{ijkt})(PCT_{ijkt})(SPT_{ijkt}) + OOT_{ijkt} \quad (4.6)$$

$$TIT_{ijkt} = DET_{ijkt} + (LRT_{ijkt})(LRR_{ijkt}) + MET_{ijkt} \\ + CET_{ijkt} + EET_{ijkt} + CRE_{ijkt} RRE_{ijkt} \quad (4.7) \\ + OFT_{ijkt} + RET_{ijkt} + OET_{ijkt}$$

$$TPT_{ijkt} = \frac{TOT_{ijkt}}{TIT_{ijkt}} \quad (4.8)$$

Partial productivity considers only one input factor. Total factor productivity considers only labor and capital input factors. For example, partial productivity with respect to labor input is expressed as follows:

$$PPT_{ijkt} \text{ (labor)} = \frac{TCI_{ijkt} + FMP_{ijkt} + PMP_{ijkt} + (FUP_{ijkt})(SPT_{ijkt}) + (PUP_{ijkt})(PCT_{ijkt})(SPT_{ijkt}) + OOT_{ijkt}}{(LRT_{ijkt})(LRR_{ijkt})} \quad (4.9)$$

Allocation criteria for overhead expenses used for output and input elements can vary depending on the type of environment, cost element, accounting information and managerial preferences (Edosomwan, 1985a, d; 1987a, b, c; Kendrick, 1984). To allocate overhead expenses to the various input components of TOTPMM, the complexity factor of technology or equipment is recommended. The complexity factor for each technology or equipment is obtained from the functional relationships of five major variables shown in expression (4.10). All variables are normalized to the same time period and ratio base.

Technology Complexity Factor (TCF)

$$\text{TCF} = f(\text{Mir}, \text{Meu}, \text{Cpt}, \text{Mur}, \text{Dhu}) \qquad (4.10)$$

where:

Mir = Machine insertion rate
Meu = Machine energy utilization rate
Cpt = Chip preparation time
Mur = Machine utilization time
Dhu = Direct hours utilized

The complexity of the technology, product, or task is used to determine a proportional contribution to the total quantity produced. Complexity in this case means items such as number of assemblies within a given technology, insertion

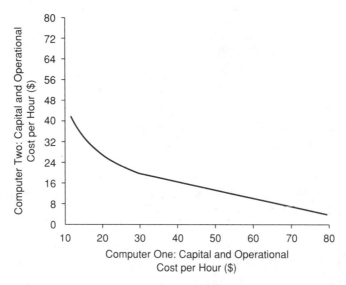

FIGURE 4.3 Typical equal productivity curve for processing 50 letters per hour.

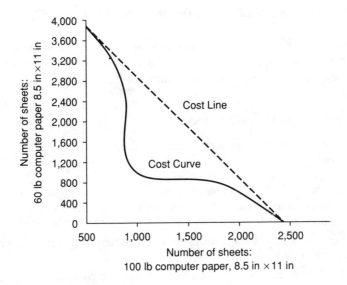

FIGURE 4.4 Typical cost curve and cost line for $80 worth of paper for a portable computer.

rate kitting time, and so on. Two commonly used criteria for allocating overhead are as follows:

Quantity Produced Proportionality Criterion. The proportional contribution to the total quantity produced allocation method is applicable in situations where there is more than one type of technology, product, or task.

Direct Hours Allocation Criterion. This approach requires daily direct labor recording and the derivation of burden rate from net expense and direct hours. Therefore, cost of product includes labor and burden. Allocation of overhead is done using direct hours and one burden rate.

4.8 EQUAL PRODUCTIVITY CURVE JUSTIFICATION TECHNIQUE

An equal productivity curve is a visual method of comparing the cost of two types of technology or equipment, each producing the same output. Figure 4.3 shows a typical equal productivity curve for processing 50 letters per hour through two types of computers.

4.9 COST CURVES JUSTIFICATION TECHNIQUE

Cost curves provide the basis for a visual comparison of the amount of materials and other supplies that can be purchased for the same sum. Figure 4.4 shows a typical cost curve and cost line for $80 worth of paper for a portable computer.

72 JUSTIFYING AND IMPLEMENTING TECHNOLOGY

4.10 SCORING MODELS FOR JUSTIFYING TECHNOLOGY

Scoring models for justifying technology take into account intangible factors that are usually not considered in profitability criteria. Unweighted scoring models and weighted scoring models have been recommended.

4.10.1 Unweighted Scoring Model

Meredit (1986) recommends an unweighted 0–1 factor model that uses the following procedure:

1. Beginning with the selection of a set of relevant evaluation factors, each technology project is rated: 1 if the technology alternative satisfies the factor, 0 if it does not.
2. The number of qualifying factors are summed for each alternative, providing a score that ranks the technology projects.
3. Technology alternative is selected based on the highest rank.

The unweighted scoring model has the advantage of minimal complexity in computation and data gathering. The disadvantages associated with the use of this model are that each of the criteria is given equal importance and each alternative either passes or fails each test. No provisions are made for indicating the degree to which factors are met. An example of unweighted scoring model application is presented in Table 4.3.

4.10.2 Weighted Scoring Model

Carrasco and Kengskool (1988) have shown how the unweighted 0–1 scoring model can be extended (see Table 4.4). Weights may be established by any method acceptable to the organization; the weights in Table 4.4 were developed by (1) ranking the decision factors in order of importance, (2) assigning the most important a weight of 1.0, and (3) assigning the remaining factors a weight based on their

TABLE 4.3 Unweighted (0–1) Scoring Model

Decision Factors	Technology Alternatives		
	1	2	3
Net present value	1	1	0
Increased productivity	0	1	1
Improved product flexibility	0	1	1
Improved quality and reliability	1	1	1
Compatibility	1	0	0
Score	3	4	3

TABLE 4.4 Weighted Scoring Model Example

Weight	Decision Factors	Technology Alternatives		
		1	2	3
1.0	Net present value	1.0	0.9	0.4
0.9	Increased productivity	0.4	1.0	0.8
0.8	Improved product flexibility	0.4	0.7	1.0
0.8	Improved quality and reliability	0.7	0.7	1.0
0.6	Compatibility	1.0	0.6	0.5
	Score[a]	2.84	3.28	3.02

[a]See equation (4.11).

relative importance compared to the first. The next step was to rank and rate the alternatives for each decision factor, again rating lesser alternatives relative to the best. The final score for each alternative is determined by summing the products of the weights and the ratings.

$$S_i = \sum_{j=1}^{n} w_j x_{ij} \qquad (4.11)$$

where:

S_i = total score for project i
w_j = weight of factor j
x_{ij} = rating of project i on factor j
n = number of factors

Several intangible factors must also be considered. These intangible factors include but are not limited to the safety of the technology for workers, the impact of the technology on employment and community relations, ergonomics and task design factors, the psychological well-being of workers, job satisfaction, the quality of working life, and the ease of workers' understanding of the new technology.

4.11 ALLOCATING RESOURCES TO DIFFERENT TECHNOLOGY CHOICES

Saad (1988) presented a linear programming model to optimize the allocation of resources to different technology choices. The notations used in Saad's model are as follows:

j = index for the chosen investment projects; $j = 1, 2, ..., n$
i = index for the funding sources; $i = 1, 2, ..., m$
π_j = average rate of return expected from alternative j

C_i = capital cost per dollar acquired from source i. It represents the interest rates on external funds (loans) and profits on internal funds (retained earnings).

u_{ij} = a weight assigned per dollar from source i used to fund project j. It represents the degree of risk associated with using source i to fund project j.

V_{ij} = the weighted value per dollar from source i, used to fund project j.

$v_{ij} = (\pi_j - C_j)U_{ij}$, for all i,j pairs. It represents an integrated measure of both profitability $(\pi_j - C_j)$ and risk (U_{ij}) of each source when used to fund each alternative.

x_{ij} = the dollar amount from source i to be used for funding alternative j

A_i = the amount of money accessible from source i

R_j = the amount of capital required to carry out alternative j

The mathematical formulation of Saad's model requires the maximizing of

From \ To			Usages				Funds Available
			T_1	T_2	T_j	T_n	
Fund Sources	Internal	Retained Profits C_1	V_{11}	V_{12}	V_{1j}	V_{1n}	A_1
		Capital addition C_2	V_{21}	V_{22}	V_{2j}	V_{2n}	A_2
		C_i	V_{i1}	V_{i2}	V_{ij}	V_{in}	A_i
	External	Long range loans C_{m-1}	$V_{(m-1)1}$	$V_{(m-1)2}$	$V_{(m-1)j}$	$V_{(m-1)n}$	$A_{(m-1)}$
		Short range loans C_m	V_{m1}	V_{m2}	V_{mj}	V_{mn}	A_m
Funds required			R_1	R_2	R_j	R_n	ΣA_i / ΣR_j

$v_{ij} = u_{ij}(\pi_j - C_i)$ $i = 1,2,...,m$, $j = 1,2,...,n$
$0 \le u_{ij} \le 1$ for all i, j

FIGURE 4.5 Model for resource allocations to different technology choices.

$$Z: \sum_i \sum_j v_{ij} x_{ij}$$

subject to

$$\sum_i x_{ij} = R_j \quad j = 1, 2, \ldots, n \quad (4.12)$$

$$\sum_j x_{ij} \leq A_i \quad i = 1, 2, \ldots, m \quad (4.13)$$

$$x_{ij} \geq 0 \quad \text{for all } i, j \quad (4.14)$$

The model seeks the optimal funding scheme that maximizes the sum of the net values (utilities) generated by all projects, subject to

1. Fulfilling the capital requirements for all the projects.
2. Not exceeding the funds available.
3. Standard nonnegativity restriction of investments.

Figure 4.5 presents the matrix coefficients corresponding to the transportation method structure. The total funds available from all sources should be

$$\sum_{i=1}^{m} A_i$$

equal to the total funds required for all projects:

$$\sum_{j=1}^{n} R_j$$

An artificial column (or row) might be added with zero profit coefficient to achieve the balance between the total funds required and available.

4.12 TEN-STEP APPROACH FOR IMPLEMENTING A NEW TECHNOLOGY

1. Perform a thorough analysis of the technology's impact on task, people, and the work organization.
2. Assess the total feasibility of the technology's original objectives, goals, or problem statement.
3. Screen the technology for relevance and appropriateness.
4. Involve all potential users of the technology in the implementation project. This is usually done by a task force across functional organizations.
5. Arrange for adequate training of the workforce on the commissioning, calibrating, and maintenance of the technology.

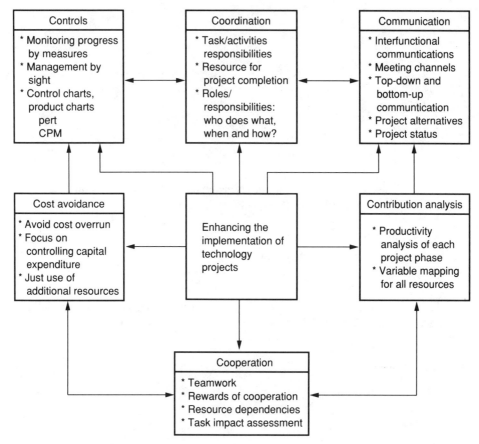

FIGURE 4.6 The 6C principles for managing technology projects.

6. Characterize the technology to understand the variability it may cause within the production or service process.
7. Demonstrate the value of the technology in social and economic terms through a pilot project.
8. Modify features or functions of the technology to suit qualification criteria or potential problems experienced.
9. Implement systems to check all input and output variables, specifications, parameters, and variables related to the technology process.

4.13 THE 6C PRINCIPLES FOR MANAGING TECHNOLOGY AND RESEARCH AND DEVELOPMENT PROJECTS

The successful management of technology projects requires the implementation and adherence to the 6C principles presented in Figure 4.6. The details of each principle will now be discussed.

4.13.1 Principle One: Provide Controls

In managing technology and R&D projects it is important to define the objectives of the project and understand the activities involved. Specific performance parameters should be used in defining the objectives. The schedules for the various activities have to be identified. Two techniques are highly recommended for controlling all the elements involved in a project. A Gantt chart provides a means for arranging the list of activities to be accomplished. Activities are arranged vertically, with the time of completion stated horizontally. The pert chart provides a means for managing the probability of timely completion for each task and highlights the earliest and latest times a given task can be accomplished.

4.13.2 Principle Two: Provide a Focal Point For Coordination

The implementation of a new technology in the workplace requires creation of a focal point for coordination of all activities. This can be achieved by designating a project manager to be in charge of all activities. The project manager, working with other members of the project team, ensures that all project resources are controlled and allocated properly, and that the project is going according to schedule. The project manager requires a good data base and tracking mechanisms for all activities. The successful project manager has good interpersonal skills, good organizational and planning abilities, and good judgment.

4.13.3 Principle Three: Provide Adequate Communication Channels

The status of a technology project must be periodically reviewed by the people involved. Usually the project manager ensures that regular meetings are held to discuss issues that require resolution and to obtain status reports on key activities. Ongoing communication among project team members is required to prevent things from falling through the cracks. Team members, vendors, and management must be involved in all project review processes.

4.13.4 Principle Four: Provide Adequate Focus on Cost Avoidance

In order to avoid cost overrun in technology project implementations, the implementation phase's cost posture should be monitored. Additional features or functions without value should be avoided. Careful attention must be paid to capital expenditures on fixtures, programming, manpower, and contractors at each phase of the project implement. In situations where vendors are responsible for project implementation, periodic status on cost performance must be requested and reviewed by the project manager.

4.13.5 Principle Five: Implement Measures to Analyze the Contribution of Each Phase

Each implementation phase of a technology project should provide tangible results. For example, the machine characterization phase should enable the project manager to understand the pattern of performance for a given machine. The contribution analysis of each phase of the technology project can be performed by using the variable-and-result mapping technique. This technique requires that for each activity or task performed, the expected result must be matched against the true output or result. This provides a quick way of identifying deviations from project goals and objectives as well as understanding the causes of deviation from specifications. This mapping technique requires that timely action be taken to resolve project deviations.

4.13.6 Principle Six: Facilitate Cooperation among Project Participants

The cooperation among members of the project implementation team is a key requirement for success. If possible, the teamwork process should involve the designer, developer, implementers, and users of the technology. When the physical presence of all the functional areas is impossible, communication channels should be put in place to facilitate information sharing that promotes cooperation. All members of the organization should understand the goals and objectives behind the implementation of a new technology in the workplace to facilitate cooperation. Such understanding can be accomplished through a project kick-off meeting at various levels of the organization.

4.14 COMMON IMPLEMENTATION PROBLEMS

Whenever a new technology, program, or idea is promoted for implementation, there are bound to be problems. The most frequent types of problems are:

1. Resistance to change from old machinery to a new type of equipment or from a manual method to a technology-assisted method. People resist change depending on how they are affected by the change.
2. Unwillingness to change work habits.
3. Fear of the unknown. People are often afraid that technology implementation might cause unemployment, decrease job satisfaction, and increase psychological stress.
4. Inappropriate vendor support for equipment specifications, documentation, and service.
5. Lack of proper project planning and monitoring by implementation team members.
6. Conflicting views of objectives and lack of clear definition of the use of technology.
7. Failure of specific equipment or technology implemented to deliver the expected result, so that additional process and procedural bottlenecks are created by the new technology.

TABLE 4.5 Project Phases, with Conflict Sources and Solution Strategies

Project Life-Cycle Phase	Conflict Source	Solution Strategy
Project formation	Priorities	Clearly defined plans, joint decision making.
	Procedures	Develop detailed administrative operating procedures. Develop statement of understanding or charter.
	Schedules	Develop schedule commitments in advance of actual project commencement. Forecast other departmental priorities and possible impact on project.
Build-up phase	Priorities	Provide effective feedback to support areas on forecasted project plans and needs via status review sessions.
	Schedules	Schedule work breakdown packages (project subunits) in cooperation with functional groups.
	Procedures	Contingency planning on key administrative issues.
Main program	Schedules	Continually monitor work in progress. Communicate results to affected parties. Forecast problems and consider alternatives. Identify potential trouble spots needing closer surveillance.
	Technical	Early resolution of technical problems. Communication of schedule and budget restraints to technical personnel. Emphasize adequate, early technical testing. Facilitate early agreement on final designs.
	Manpower	Forecast and communicate manpower requirements early. Establish manpower requirements and priorities with functional and staff groups.

TABLE 4.5 (Continued)

Project Life-Cycle Phase	Conflict Source	Solution Strategy
Phase-out	Schedules	Close schedule monitoring in project life cycle Consider reallocation of available manpower to critical project areas prone to schedule slippages. Attain prompt resolution of technical issues that may impact schedules.
	Personality and Manpower	Develop plans for reallocation of manpower upon project completion. Maintain harmonious working relationships with project team and support groups. Try to loosen up "high-stress" environment.

8. Overcomplexity. The new tool or equipment is too complex for the work organization; no adequate support structure exists for the new equipment or tool.

Croft and Ledbetter (1988) recommend four phases for managing project life cycle: project formation, build-up phase, main program, and phase-out. Details of these phases are presented in Table 4.5. Gaynor (1988) recommends that the following questions be asked and answered when evaluating projects:

1. Are the objectives of the technology project clearly defined?
2. Do the objectives of the project support the strategies of the organization?
3. Why is this project important?
4. Is this a business project or technology project?
5. Could the research effort, for example, be reduced or minimized by joining forces with other organizations?
6. Does the project leverage the current organizational bank of technology? Is the technology understood by the organization? Can the benefits of that technology be clearly delineated? What is the "value added" in terms of new knowledge?
7. Is the technology proprietary, that is, covered by patents? What other technologies could competition develop to achieve the same results? Has the applicable technology been explored on a global basis?
8. Can the known and unknown aspects of the technologies be clearly stated? What are the specific technologies involved?

9. If the project is related to research, what are the estimated costs to the organization (i.e., the total cost)? What are the potential capital requirements if successful? Is the sales and marketing organization in place as well as the total distribution system? If not, what are the alternatives? How would the potential product or service be introduced to the public? What are the potential product liabilities? What is the impact of possible government intervention?
10. Has a thorough analysis been made regarding the project staffing? Are details of the staff needs already developed? Is the staff available internally or will additions of new personnel be required? Is the talent available from the outside and will it be necessary to bid against other users?
11. Can the required expertise and the level of expertise be defined? Are the basic required competencies already documented; that is, not can a particular individual probably do the job but does the competency exist and has it been demonstrated reliably and effectively in the past? Do the proposed individuals have a track record of successes?
12. Is there a project manager with the credentials plus the experience and the track record? Have the requirements been detailed? (The requirements are quite different for research, product development, process, manufacturing, engineering, business, etc.) What expertise should the individual bring to the project? If required, can he manage interfaces? Can the person function in a matrix organization? Does the manager bring a balanced perspective to the project? Is the individual business oriented? Is the individual parochial, or will he/she fully utilize not only internal expertise but outside resources as well? Does the individual meet commitments? Can the selected manager function in an interdisciplinary and multifunctional environment? Can the person instill the essential discipline of meeting commitments and at the same time maintain an environment that stimulates innovation, dedication, and strives for excellence?

4.15 OVERALL STRATEGY FOR OVERCOMING COMMON PROBLEMS

The following eight actions are recommended for implementing a new technology in the work environment:

1. Obtain the total support of the total organization and the vendors involved in the technology implementation project.
2. Set realistic implementation goals and objectives that maximize the effective use of resources.
3. Plan implementation activities ahead of time.
4. Maintain internal and external contacts with key people who have the expertise to help. Do not reinvent the wheel.

5. Find out who has implemented such equipment, tools, or technology before. Learn from their experiences.
6. Follow up and monitor project activities closely. Close up open technical issues quickly and effectively.
7. Sell the benefits of the new technology effectively to the total organization. Show implementation progress to everyone involved.
8. Keep a positive team attitude that the implementation of the new technology will produce successful results.

Gaynor (1988) recommends the following guidelines to ensure successful technology project management:

1. It is essential that all participants understand their role, the expectations regarding their performance, and how their particular contribution fits into the total scheme. Too often assignments are made without providing project staff with an overall understanding of the objectives or of where the program fits into the corporate scheme.
2. Focus attention on the key factors as well as the critical problems that would preempt success. Technical problems that cannot be solved at the laboratory bench or cannot be demonstrated in experimental models are seldom solved on the plant floor. Monitoring of specific problem solutions and understanding the limitations or potential inconsistencies of those solutions require continual evaluation.
3. Project reviews should be a generic part of any project, but how and in what depth they are used to focus on problems determines their value. Too often project reviews are limited to a discussion of cost, schedule, and performance. If technical specialists are introduced, the examination of approaches is often limited and is generally too late since certain decisions have already established the direction. To be effective, project reviews must be in depth, must be directed at a full understanding of proposed solutions, and above all require time. The sooner the project manager institutes project reviews with an understanding that integrity during discussions is primary, the better chance there is for success.

During the life of a project, many variations will be introduced. These changes could be related to the business, the marketplace, or the technologies. Reviewing the assumptions on which the project was originally approved is a continuous process. Too often, not until the completion of a project do people recognize that the market no longer exists, the technology is obsolete, or the original purposes no longer fit the current business strategy.

Project reviews should focus on answering difficult questions such as the following:

1. What key factors, technical as well as business, will preclude this project from being successful?
2. Have the contingencies been given sufficient consideration? Will the project be on schedule until the last two months and then be a year late?
3. Are project plans updated on a regular time schedule?
4. Is project monitoring being performed against the plan?
5. Are assigned personnel doing what they said they would do? Are objectives being met in a timely manner?
6. Do the current approaches or designs meet the project requirements?
7. Have reliability and quality been considered? Are new waste targets proposed or will status quo targets prevail?
8. What are the potential problems in the manufacturing or implementation start-up phase?
9. If delays are encountered that go beyond the allotted time, can additional resources be brought in to fill the gap?
10. Have the critical problems been solved, or only apparently solved, and do solutions have adequate documentation?
11. Are solutions reviewed by independent specialists who observe the performance as well as read the reports?
12. Are reports written to please management or to provide an accurate picture of the project status?
13. Are reports sent only up the ladder or are all project personnel informed as to the status of the project?
14. Are delegated activities followed up to verify performance?
15. Are new critical issues, arising from newly gained knowledge, given adequate consideration?
16. What has the competition been doing in the interim?
17. Is the project staffed with the required levels of talent? Have the requirements changed? Should new personnel be brought into the project?
18. Are all available resources being used?
19. Is there a continual focus on trying to accelerate the project completion?
20. Who should attend review meetings? Are you depending on your supervision or are you asking the "doers" to report on their progress?

Croft and Ledbetter (1988) offer the following guidelines for managing technology projects:

1. Establish a central coordination point for all services provided.
2. Provide integrated and timely planning.
3. Provide for responsive service to customers.

4. Permit the shifting of resources to accommodate fluctuating workloads.
5. Establish within one organization the information as to where, when, why, and how resources are being expanded.
6. Provide management with performance visibility.
7. Eliminate redundancies in administrative work control.

It is also important to ensure that the following activities are performed:

1. The project has been subdivided into manageable work tasks, called *work packages*, of relatively short duration and small cost.
2. Each work package has been assigned a manager fully responsible for it.
3. A separate cost account has been established for each work package.
4. Each work package has been defined by a concise statement of the work, with functional and performance characteristics defined.
5. A schedule and corresponding budget have been established for each work package.
6. Clear start and stop dates have been defined.

Meredith and Green (1988) have reviewed the literature and offer 10 recommendations to assist managers interested in implementing new technologies:

1. Do not try to sneak up on your employees.
2. While the technology is changing, change the organization.
3. Embrace the negative consequences.
4. With more computerized communication, increase the amount of face-to-face communication.
5. Attack the strongest for stability, the existing infrastructure.
6. Prepare to fight the same old problems.
7. Keep the production utilization of the new equipment low.
8. Do not expect to eliminate jobs.
9. Use the new technology to monitor your workers.
10. Change your strategy as you go.

Meredith and Green also offer some guidance and warnings for firms working to introduce new technology:

1. Be clear in your own mind before implementing what you are trying to accomplish strategically with the new technology. This may help you anticipate how you want to manage the implementation process. If certain negative consequences appear inevitable, given your strategy, move to accomplish those as soon as possible and move on; uncertainty breeds rumors and dissipates the possible benefits of new technology.

2. Assess the functional characteristics of a new technology and attempt to anticipate the "unavoidable" organizational consequences of it. Embrace these consequences and take action to capitalize on them as opposed to trying to overcome or just endure them.
3. Treat the disease, not the symptoms. The firm needs to recognize that the individual and group consequences of new technology are symptoms of fundamental changes to the organization's form and operation. To attack the negative consequences of new technology, management must address the organizational consequences of the technology and intervene as appropriate.
4. Expand your view of "the job." New technology offers many trade-offs. Recognize that the old "job" may appear to be eliminated by the technology, but what appears to be elimination may be transformation. A rich individual or manual skills job may be transformed into a richer group or cognitive skills job.
5. Be prepared to change your strategy. In a number of cases, new technology brought unexpected benefits that allowed the firm to change from a defensive posture to a growth posture. Be prepared to recognize these opportunities by constantly assessing the state of your competitive environment and the strategic implications of the technology.
6. Stay flexible and responsive to both problems and opportunities with the new technology. Monitor the organizational and group consequences and take action to address the issues that will inevitably arise.

Successful justification and implementation of new technology demand teamwork and excellent performance from all project participants. Periodic evaluation of all activities and performance is recommended.

4.16 MEASURING TECHNOLOGICAL PROGRESS

Let the production function of a specific technology output of a factory be represented by the expression

$$P = T(t)f(L,C,M,E) \tag{4.15}$$

Where $T(t)$ is a function of all the factors that go into determining P besides C (machine-hours), L (labor-hours), material (M), and energy (E); that is, changes in T provide a way of estimating the relative importance of technical progress (TP) in the growth of output.

$$\frac{dP}{dt} = \frac{dT}{dt}f(L,C,M,E) + T\frac{df(L,C,M,E)}{dt} \tag{4.16}$$

$$\frac{dP}{dt} = \frac{dT}{dt} \cdot \frac{P}{T} + \frac{P}{f(L,C,M,E)}[\frac{\partial f}{\partial L} \cdot \frac{dL}{dt} + \frac{\partial f}{\partial c} \cdot \frac{dC}{dt} + \frac{\partial f}{\partial M} \cdot \frac{dM}{dt} + \frac{\partial f}{\partial E} \cdot \frac{dE}{dt}]$$ (4.17)

Dividing equation (4.17) by P:

$$\frac{dP/dt}{P} = \frac{dT/dt}{T} + \frac{\partial f/\partial L}{f(L,C,M,E)} \cdot \frac{dL}{dt} + \frac{\partial f/\partial c}{f(L,C,M,E)} \cdot \frac{dC}{dt} + \frac{\partial f/\partial M}{f(L,C,M,E)} \cdot \frac{dM}{dt} + \frac{\partial f/\partial E}{f(L,C,M,E)} \cdot \frac{dE}{dt}$$ (4.18)

For any variable Z, $(dZ/dt)/Z$ is the geometric rate of growth of Z per unit time, called GZ. Hence, equation (4,18) can be written as

$$GP = GT + \frac{\partial f}{\partial L} \cdot \frac{L}{f(L,C,M,E)} \cdot GL + \frac{\partial f}{\partial c} \cdot \frac{C}{f(L,C,M,E)} \cdot GC + \frac{\partial f}{\partial M} \cdot \frac{M}{f(L,C,M,E)} \cdot Gm + \frac{\partial f}{\partial E} \cdot \frac{E}{F(L,C,M,E)} \cdot GE$$ (4.19)

Each term on the right-hand side of equation (4.19) now becomes an elasticity:

$$\frac{\partial f}{\partial L} \cdot \frac{L}{f(L,C,M,E)} = \frac{\partial P}{L} \cdot \frac{L}{P} = \text{elasticity of output with respect to labor input,} \quad Ep,L \quad (4.20)$$

$$\frac{\partial f}{\partial C} \cdot \frac{C}{f(L,C,M,E)} = \frac{\partial P}{C} \cdot \frac{C}{P} = \text{elasticity of output with respect to capital input,} \quad Ep,C \quad (4.21)$$

$$\frac{\partial f}{\partial M} \cdot \frac{M}{f(L,C,M,E)} = \frac{\partial P}{M} \cdot \frac{M}{P} = \text{elasticity of output with respect to material input,} \quad Ep,M \quad (4.22)$$

$$\frac{\partial f}{E} \cdot \frac{E}{f(L,C,M,E)} = \frac{\partial P}{E} \cdot \frac{E}{P} = \text{elasticity of output with respect to Energy input,} \quad Ep,E \quad (4.23)$$

Equation (4.18) now becomes

$$GP = GT + (Ep, L)GL + (Ep, C)GC + (Ep, M)GM + (Ep, E)GE \quad (4.24)$$

and GT is the relative importance of technological progress in determining growth of output:

$$GT = GP - (Ep, L)GL - (Ep, C)GC - (Ep, M)GM - (Ep, E)GE \quad (4.25)$$

As an example, assume that a certain computer factory had the following data for 1984–1988:

$GP = 3.95$ % per year
$GL = 1.25$ % per year
$GC = 1.35$ % per year
$GE = 1.00$ % per year
$GE = 0.75$ % per year
Ep, $L = 0.48$
Ep, $C = 0.60$
Ep, $M = 0.51$
Ep, $E = 0.42$

With these data, GT can be computed:

$$GT = 3.95 - 0.48(1.25) - 0.60(1.35) - 0.51(1.00) - 0.42(.075)$$
$$= 3.95 - 0.6 - 0.81 - 0.51 - 0.32$$
$$= 1.71$$

Technology at the computer factory advanced at rate of 1.71% per year over the period 1984–1988.

The technological progress factor $T(t)$ can enter into the production function in two ways.

1. Neutral technical progress: In this situation technological progress affects all inputs equally, and equation (4.15) is used.
2. Individual input augmenting technological progress: for example, with respect to capital (machine related) input,

$$P = fT(t), C, L, M, E$$

Each of the two types of technological progress has the effect of shifting the production function. Over time, more output can be obtained from any given combination of inputs.

SUMMARY

The introduction of technology or equipment into the workplace involves substantial capital expenditures. A total systems approach should be used to justify new technologies. This technique captures the total cost of doing business and the total output based on the expected use of the new technology. Other financial justification techniques are also useful, not as a stand-alone methodology but as a sanity check. The comprehensive justification of a new technology requires that managerial judgment be applied to weigh both the tangible and intangible benefits and costs associated with the use of the new technology.

The implementation of a new technology in the workplace is perhaps one of the most interesting and challenging experiences for decision makers. Most projects fail either because they have not been justified properly or because the ingredients and mechanisms for implementation were not carried out properly. The teamwork approach among functional areas, vendors, developers, and users of technology is highly recommended for a technology project implementation. The comprehensive implementation of the 6C principles of control, coordination, communication, cost avoidance, contribution analysis, and cooperation provides a strong base for the successful implementation of a new technology in the workplace.

QUESTIONS AND PROBLEMS

4-1 What are the disadvantages involved in using the rate of return on investment technique to justify a new technology?

4-2 Discuss the uncertainties involved in making capital investment decisions. How would you handle these various uncertainties?

4-3 What are the advantages and disadvantages of using the technology-oriented total productivity measurement model (TOTPMM) as a justification technique for technology-based systems?

4-4 Describe the application of the cost curve and the equal productivity curve.

4-5 What are the limitations of the payback period method?

4-6 Discuss the various intangible factors that should be considered in technology justification.

4-7 Describe a step-by-step methodology for justifying a new tool or equipment.

4-8 How would you sell the implementation of a new technology to top management of a medium-size firm?

4-9 Discuss the various depreciation methods for capital equipment. What are their advantages and disadvantages?

4-10 Suggest a weighting scheme for the approximation of intangible factors in technology justification. How can tangible and intangible factors be reduced to the same unit of measurement?

4-11 What are the roles of a project manager in implementing technology-based systems?

4-12 Describe a step-by-step approach for implementing the 6C principles.

4-13 What types of controls are necessary for technology projects? How would you implement such controls?

4-14 How would you facilitate the coordination of technology projects during the design, production, and implementation phases?

4-15 Describe four techniques for cost avoidance.

4-16 Assume you are the analyst responsible for implementing a new insertion machine for the production of sport shoes. What controls will you put in place to monitor the progress of the project?

4-17 Develop a strategy for handling vendor interactions in technology projects.

4-18 Discuss the common technology implementation problems.

4-19 What are the benefits of having a project office responsible for technology projects? How would the concept of a project office be perceived by line management?

4-20 How would you go about developing and implementing a team approach for managing high-technology projects?

5
MANAGING TECHNOLOGY UTILIZATION, TRANSFER, AND FORECASTING

This chapter discusses approaches for managing technology utilization, transfer, and forecasting. User-oriented frameworks of technology are presented.

5.1 MANAGING TECHNOLOGY UTILIZATION

As shown in Figure 5.1, the Technology User Problem Solver (TUPS) framework divides technology utilization problems into four stages: (1) technology design and development, (2) technology cost and performance evaluation, (3) training personnel for technology utilization, and (4) technology maintenance and repair.

The minimization of technology utilization problems starts with good design. At the technology and development stage, systems designers are urged to use the comprehensive design rules discussed in Chapter 3 to ensure that all technology-based systems designed for human use are safe and fit for use in the workplace.

The value engineering approach should be used to reduce the cost of technology to users. The value engineering approach seeks to minimize the cost of every component or feature involved in a technology-based system. Often additional parts and features are purchased for technology-based systems with no understanding of whether there is optimum use for such items. When deciding on additional requirements, features, or functions, decision makers should use the following four-step approach.

1. Clearly identify the need for extra features, functions, or options.
2. Identify the function and its cost effectiveness.

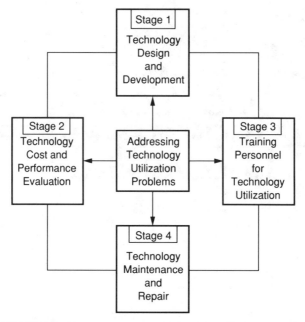

FIGURE 5.1 Technology user problem solver (TUPS) framework

3. Evaluate the worth and productivity of the additional feature or function by comparison.
4. Develop value alternatives and obtain additional features, functions, or options if needed.

Training of the workforce on how to operate and maintain the technology-based system once it has been installed is of great importance. Training services should be geared toward providing the workforce with adequate background for commissioning, calibrating, operating, and maintaining and repairing technology, no matter what the size or scope. Redefinition of a new task sequence for the worker will also be necessary. Retraining for new jobs will have to be considered for the displaced workers.

Technology utilization problems can be greatly minimized if appropriate maintenance and repair centers are set up and geared toward providing spare parts, qualified technicians, and working drawings of equipment. Often, the lack of repair and maintenance facilities has led to machine downtime and subsequent decline in labor and capital productivity. This is especially the case in developing countries, where there are no qualified personnel to repair equipment and it sometimes takes several months to receive spare parts from suppliers of equipment. One way to minimize repair and maintenance problems is through a joint effort of suppliers and users of technology in setting up maintenance centers that provide the technical assistance, spare parts, and equipment information. Such centers can also help in adapting technology transferred from other countries for local use. A preventive maintenance program is also recommended for existing technologies.

To evaluate the effectiveness of such specific preventive maintenance programs, order techniques can be used. Corder's (1971) formula for evaluating the efficiency of a specific maintenance program is as follows:

$$E = \frac{K}{C + L + W}$$

where

E = efficiency index
K = a constant equating E to 100 in the base year
C = maintenance, labor, and material cost − equipment replacement cost (these factors may be weighted to balance cash values
L = downtime − scheduled operating time
W = percentage of scheduled production wasted because of maintenance

Technology assessment must be a full-time, fully funded and directed effort on the part of the source of the technology. Technology utilization requires considerable liaison among the people who develop the ideas, their agents, and the people who originate the concepts. Sometimes they will need assistance from other sources of expertise. Person-to-person contact is necessary.

5.2 MANAGING TECHNOLOGY TRANSFER

In a broad sense, technology transfer can be defined as a process by which a body of technical knowledge, techniques, or tools is transferred from one place to another for use in a production or service environment. The management of technology transfer issues from one location to another is not an easy task. Transfer agents for technology must meet the restrictive patent and trade policies of other organizations and countries. The successful management of the transfer of technology requires a joint effort among the developer, marketer, and user of technology to foster relationships that permit the right level of training, the required approvals, and person-to-person contact at the firm and government levels.

5.2.1 Obstacles to Technology Transfer

According to Mogavero and Shane (1982), the universal resistance to change, both rational and irrational, and restrictive patent and trade policies are factors common to both developed and developing countries. They identify the major barriers to technology transfer as:

1. Inadequate skills in the entire gamut of functions from management to semiskilled labor.

2. Inadequate producer-consumer relationships.
3. Inadequate physical, social, and economic infrastructure. While possibly not inadequate in the local culture, it is often not structured for new technologies.
4. Scanty knowledge of local resources such as water, soils, or raw materials.
5. Lack of capital and incentives for investment.
6. Concentration of resource ownership in a minute sector of the population.
7. Restricted mobility of labor.
8. Lack of existent appropriate technologies for local conditions. As an example, advanced industrial nations normally develop technology that trades off relatively large amounts of capital to reduce labor requirements. Such technologies may not be appropriate to conditions found in developing countries.

5.2.2 Methods of Technology Transfer

There are five major avenues for transferring technology from one place to another:

1. The idea for a new technology can be transferred through the process of technical information exchange. Technical papers published in journals and proceedings can be a viable source for cross-fertilization of ideas in technology issues and other operating systems. Technologists worldwide have been known to transfer vital technical information that often leads to new breakthroughs.
2. Technology may be transferred through a disjointed process that involves purchasing specific technology or technical information. This channel relies heavily on local expertise to transfer the new technology and information for local use. It places a greater burden of risk on the purchaser, who may not have the adequate sources of help when problems arise with the acquired packages.
3. A purchaser with enough seed capital may acquire a complete, ready-made system, including the service agreements. This avenue usually involves huge sums of capital. The burden of maintenance falls on the owner when the service contract expires, after which the required spare parts and technical expertise are often unavailable.
4. The fourth avenue for transferring technology is through international agreement between governments, multinational corporations, firms, and individuals. This usually involves exchanging key resources, such as oil and minerals, for specific technologies. The donor often gives technologies that are antiquated, tailor-made for developed economies, and not proprietary. Again, the burden of maintenance and repairs is on the shoulders of the recipient, who often has no adequate facilities or trained personnel to maintain such technologies.
5. The fifth avenue for transferring technology is through a process called *indigenous technology adaptation process*. The purchaser and seller of

the technology information reside in similar economic environments, and adaptation strategies for the technologies are in place. This avenue is perhaps the best since it takes into account, the need, requirements, and expertise available locally.

Eldin (1988) provides the following guidelines for foreign executives involved in initiating and planning of technology transfer.

1. Be aware of the environment of the host country.
2. Study the host country in developing plans and develop a long-range plan for the future cooperation.
3. Survey the financial institutions involved in the project and study investment loans if applicable.
4. Before undertaking a project, study its feasibility from the technical, economic, and operational point of view. Also study the contribution of the development of the host country.
5. Develop relationships with government and business leaders and develop a special relationship with the project's local "godfather."
6. Choose the right project manager to implement the project and assign him to the project at early stage.

The following guidelines are also recommended for professionals involved in managing and implementing technology transfer projects:

1. Acquire sufficient knowledge about the new operating environment; that is, the threats and opportunities that exist in the economic, political, social, and cultural structure.
2. Understand the operating policies and decision-making process of the government and the different organizations responsible for project implementation and execution.
3. Know the individuals responsible for the technology transfer project very well. Fraudulent and corrupt proposals by individuals are sources of potential trouble.
4. Map out a comprehensive strategy to address the different phases of the project: preparation and initiation, implementation, and operation.
5. Have a contingency plan available for investment decisions.
6. Beware of high-pressure tactics from clients about incentives (taxes, subsidies, balance of payments) and long-term gains.

The degree of success in transferring technology from one location to another depends on the type of technology involved, the techniques and methods employed for transfer, the purchaser-seller relationship, the organization of the work involved in the entire technology transfer project, policy issues, and skills of the technical personnel involved in the transfer process. The following mechanisms are, therefore, recommended for the transfer of technology.

1. The training of transfer agents and users of technology.
2. The establishment of technology transfer centers and maintenance centers.
3. The establishment of training and research centers for existing and emerging technologies.
4. Availability of capital to purchase tools, machinery embodying needed technologies, technology licenses, patents, and technology-based systems codes.
5. The ongoing maintenance of suitable environments for technological innovation and adaptability.
6. Adequate procedure for assessing the real need, screening technologies, justifying and modifying technologies, and diffusing appropriate technologies to the required sectors of the economy.

5.3 MANAGING TECHNOLOGICAL FORECASTING

It is important for the organization to have a formalized program for technological forecasting. Such a program can

1. Help to establish the timing of new technology and maximize gain from events that result from action taken by the corporation.
2. Set quantitative performance standards for new products, processes, and materials.
3. Assist in the planning of research programs: the amount and direction, the scientific skills needed, and resource utilization.
4. Guide engineering programs toward the use of new technology and the adjustment to new technical demand.
5. Help in new product development and current product improvement.
6. Help to identify major opportunities and threats in the technological environment and their social impact on employment, skills, and educational needs, for example.
7. Help to identify the economic potential and impact of technological progress, and to guide long-range planning.
8. Help to determine leadership in industry and increase the future market share and profit of the corporation.

5.3.1 Definition of Technological Forecasting

Technological forecasting is a tool used for the prediction and estimation of the feasible or desirable parameters in future technologies. In high-technology industry, technological forecasting is a prediction of what might happen given certain assumptions and objectives.

5.3.2 Operational Definition of Technological Forecasting by Levels in High-Technology Industry

At the *policy* planning level, technological forecasting is the clarification of scientific technological elements determining the future boundary conditions for corporate development. At the *strategic* planning level, technological forecasting is the recognition and comparative evaluation of alternative technological options. At the *operational*, or *tactical*, planning level, technological forecasting is the probabilistic assessment of future technology. At the *marketing* and *corporate profit* planning level, technological forecasting is the clarification of scientific technology needed to expand the market share and compete in the world market.

5.3.3 Technological Forecasting Methodology in High-Technology Industry

Owing to the great competition that exists within high-technology industry, a logical framework is needed to perform adequate forecasts. Technological forecasting methodology is presented schematically in Figure 5.2, which will now be described in detail.

Where Technological Forecasting Begins. Technological forecasting begins with a strategy meeting of the management team. The purpose of the strategy meeting is to reach an understanding of the fundamentals of the development of the various subforecasts, to assign responsibilities, and to set a schedule. The fundamentals include:

1. A review of all corporate long-term and short-term objectives.
2. An assessment of the present status and a relative forecast of the future technology needed for the total business environment.
3. A statement of the objectives and motivation for the technological forecast period within the constraints and opportunities recognized in the total business environment.

Technological forecasting may also begin with individual or group intuition. This usually occurs, after on-the-job training, in individuals with self-initiative and curiosity about what may happen in the future.

Technological Forecasting Group (TFG) Activities. Most major corporations in high-technology industry have dedicated people who have been given the responsibility for technological forecasting. The technological forecasting group is made up of various experts with interdisciplinary backgrounds. The TFG ensures that all technical assumptions and motivations for the technology are clearly defined; it

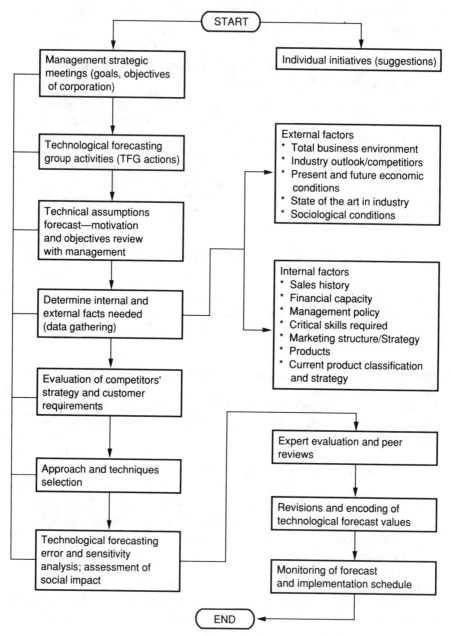

FIGURE 5.2 Technological forecasting methodology in high-technology industry.

determines the internal and external facts needed, evaluates competitor strategy and customer requirements, selects approaches and techniques to perform the forecast, assesses the social impact of the forecast, encodes variables and parameters, and monitors the technological forecast and implementation schedule.

Important Aspects of Technological Forecasting. Some important internal and external factors governing technological forecasts are shown in Figure 5.2. Data must be gathered from whichever of these areas are relevant to the forecast at hand.

Evaluation of Competitor Strategy, Customer Requirements, and the State of the Art in Specific Areas of Interest. Some of the specific questions to be answered in evaluating competitor strategy, customer requirements, and the state of the art in areas of interest are as follows:

1. What is available in high-technology industry, and what are the areas of interest?
2. What are the competitors doing?
3. What products satisfy the customer of today, and what products will satisfy the customer of tomorrow?
4. Is there consistency in such areas as manufacturing capability, market structure, and labor skills?
5. What are the latest measuring techniques (scientific papers and research efforts)?

Technological Forecast Timing. In high-technology industry the range of the technological forecast period is usually dictated by the purpose of the forecast. The technological forecast horizon must cover the cumulative lead time required for executing the plans. Technological forecasts can be classified as:

Short-term: less than 2 years.
Intermediate: 3–4 years.
Long-term: 5–20 years.

The frequency of technological forecasting and review is usually in intervals that are no longer (and usually shorter) than the length of the forecast period. For example, annual forecasts are updated quarterly, long-term forecasts annually.

Approaches and Techniques of Measurement. The two approaches usually used for technological forecasting are (1) the exploratory approach (push approach)—project the area of technological forecasting to the future—and (2) the normative approach (pull approach)—start with goals and objectives, then identify the end item or result.

In high-technology industry, statistical techniques, operation research techniques, and reliability analysis are heavily used in technological forecasting.

Single-trend Extrapolation. This technique involves extrapolating the trend of certain technological parameters; for example, the memory capacity of a personal computer, maximum aircraft speed, and operating energy of particle accelerators are plotted over long periods as a function of time. It is then assumed that the nature of the progress experienced in the past will continue to occur in the future.

Growth Analogy. Growth analogy techniques assume that biologic growth provides a useful analogy for many technologies. For example, Lenz (1968) applies the biologic growth analogy to a projection of trends in the maximum speed of military aircraft.

Regression and Correlation. This is a statistical technique used in technological forecasting to forecast the values of certain unknown parameters given historical trends. The two most popular regression methods are simple linear regression and double linear regression.

Substitution. Instead of measuring the increase in performance occurring in technology, the analyst measures the rate at which one technology is substituted for another in general usage.

Fitted Curves. This technique involves the extrapolation of time-series-related trends. The fitted curves are used as an aid to forecast what could happen in the future.

Delphi. The Delphi technique is used to obtain a consensus of experts. This technique is designed to systematically combine individual judgments and thus obtain a reasoned consensus about the technological forecasting input needed. Questionnaires are used in most cases to obtain opinions.

Scenarios. Scenarios can be either direct extrapolations of the present conditions or variations formed by adding new conditions to the present environment. The three methods of creating scenarios are the consensus technique, the cross-impact matrix, and the alteration-through-synopsis. The cross-impact matrix, which permits the orderly investigation of the effects of potential interactions among items in a forecasted set of occurrences, is the most widely used.

Expert Evaluation of Technological Forecasts. Expert evaluation is a sanity check on the technological forecast group output. The procedure ensures

1. Consistency of technological forecasting objectives with corporate strategic direction: Why is this forecast needed?
2. Appropriate technological forecasting assumptions and intelligently interpreted facts.
3. Appropriate information and data and intelligently interpreted facts.
4. Validity of the approach and the techniques used.

5. Comprehensibility of study.
6. Assessment of social impact.
7. Recognition of the limitations of the technological forecast.
8. Monitoring tools and techniques.
9. Implementation schedule and recognition of pitfalls.
10. Validity of the time horizon forecast.

Encoding of Variables, Parameters, and Values. One of the most important tasks of the technological forecast group is the coding of forecast variables, parameters, and values. This is done to:

1. Prevent potential competitors from obtaining the technological forecast information, thus avoiding a loss of the market share or product leadership.
2. Prevent unqualified practitioners from misusing the technological forecast information, especially in such areas as defense systems and intelligence monitoring.
3. Protect the product announcement schedule.
4. Minimize the loss from internal and external data sales crime.

Monitoring the Technological Forecast. Technological forecast measuring tools and the units of measure should be selected or designed to fit the subject, the product, the environment, the objective, and the user. Some examples of practical measuring tools for monitoring the status of technological forecast errors in high-technology industry are (1) the coefficient of variation (= standard deviation divided by the average projected forecast value [demand]), and the standard error (= standard deviation divided by $5N$, where N = number of observations); (2) volume index graph, line-average graph, and moving-average graph; (3) comparison statements; and (4) the company profit plan.

5.3.4 Problems and Pitfalls in Technological Forecasting

1. Inappropriate environmental study (gathering data on competitors and politics).
2. Lack of adequate data on industry averages in many technical specification areas.
3. Distortion of the forecast by strong emphasis on certain areas, such as specification or reliability information.
4. Poor technological forecast monitoring tools and techniques.
5. Lack of total involvement from the management team and technological forecasting group members.
6. Lack of documentation.

5.3.5 Keys to the Successful Use of the Technological Forecast

1. Understanding the basic assumptions of technological forecasts and their limitations; preparation to be flexible enough to manage the forecast successfully within an agreed-upon and preplanned tolerance range of forecast error.
2. Timely measurements of technological forecast errors.
3. Good management interpretation and understanding of differences between the technological forecasts and actual values.
4. Management response to the "vital signs": timely recognition and understanding of developing problems and adjustment of forecasted values and parameters to suit prevailing conditions.
5. Total involvement of all the "players."

5.3.6 Key Issues in Technological Forecasting

Technological forecasting is important because it defines our best estimate of what could happen: the opportunities and restraints of our future. Technological forecasting is an input to the planning process; it is not a plan in itself. Usually all the subsequent planning of production, materials, and capacity will be developed within the scope and parameters of a corporate technological forecast. Forecasters should begin with the following steps:

1. Identify all information that will be essential to the forecast.
2. Calculate any essential data that may be calculated with precision.
3. Forecast the balance of essential information that is required but cannot be calculated.

Good technological forecasting is accomplished by progressively gathering appropriate essential facts, interpreting the facts intelligently, modifying statistics with managerial judgment, monitoring the technological forecast error, and progressively updating and refining the forecasts.

Technological forecasting requires the involvement and commitment of every member of the forecasting group and management team. All assumptions on which technological forecasts are based should be defined, stated, understood, and agreed to by all members of the forecasting group and management team. The risk-reward relationships involved in hedging against the margin of technological forecast error should be discussed openly, to avoid conflicts and to build credibility.

Understanding and using the technological forecast is even more important management work than making the forecast. Decision makers are concerned with validity, credibility, comprehensibility, and accuracy. Attention should be paid to significant differences between the technological forecast and the actual situation; these are usually early warning signals calling for corrective action.

Technological forecast accuracy depends on many factors, including the qualitative and quantitative techniques used, the source of the data, forecaster know-how, a clear understanding of the business environment, and knowledge of the state of the art in the field.

SUMMARY

The management of technology utilization, transfer, and forecasting requires that people have access to a broadly based body of technical information and experience and that they have the freedom and motivation to seek opportunities aggressively and respond quickly to calls for assistance. The early involvement of the designers, developers, and users of technology to address utilization and transfer problems is a good strategy.

The complexities involved in technology transfer can be minimized by improving the overall quality of technical knowledge to decision makers regarding all types of technologies. Managing technology transfer from one location to another is not an easy task. Transfer agents for technology are bound to meet with restrictive patent and trade policies across organizations and countries. The successful management of the transfer of technology requires a joint effort between the developer, marketer, and user of technology to foster relationships that permit the right level of training, required approvals, and person-to-person contact at the firm and government levels. Other important considerations are:

1. The training of transfer agents and the users of technology.
2. Availability of sufficient capital to purchase tools, machinery embodying needed technologies, purchase of technology licenses, patents, and technology based systems codes.
3. The ongoing maintenance of suitable environments for technological innovation and adaptability.
4. The establishment of technology transfer centers and maintenance centers for new and existing technologies.
5. The establishment of training and research centers for existing and emerging technologies.

The quality of technological forecasting can be improved by providing adequate resources for generating more pertinent data on all relevant variables and parameters.

QUESTIONS AND PROBLEMS

5-1 Explain the problems of technology transfer in a developing economy.

5-2 Why is it easier to transfer technology in a developed economy?

5-3 Define technological forecasting.

5-4 How would you forecast a technology for a specific requirement in a computer industry?

5-5 What procedures would you follow in managing technology utilization problems?

5-6 How would you go about performing a competitive analysis on a typical technology-based system?

5-7 Discuss the various technological forecasting techniques with emphasis on their advantages and disadvantages.

5-8 What are the differences between technological forecasting and planning?

5-9 How would you use the Delphi technique in technological forecasting?

5-10 Discuss the common pitfalls in technological forecasting.

5-11 Define the roles of technology transfer agents. What are some of the difficulties that could be experienced by technology transfer agents in cross-industry technology transfer?

5-12 What mechanisms should be used to encourage technology transfer between individuals at the enterprise level?

5-13 How should the maintenance problems involved in technology transfer to developing countries be handled?

5-14 What are the specific roles that could be provided by senior managers of firms, government agencies, and public officials in promoting technology transfer?

6
MANAGING PRODUCT DESIGN AND DEVELOPMENT ACTIVITIES

This chapter discusses the characteristics of research and development (R&D) organizations. The factors that can be controlled by management and those that cannot are identified. A matrix organization structure is suggested for R&D projects. A step-by-step approach for implementing design for automation concepts and the concept of early manufacturing involvement (EMI) in the design of a new technology is provided. Guidelines are offered for designing products for automation.

6.1 CHARACTERISTICS OF RESEARCH AND DEVELOPMENT ORGANIZATIONS

An analysis of the characteristics of research and development organizations offers a number of possibilities for obtaining insight into the evolution of organizational forms in which innovation can develop more effectively. Although there are many ways of thinking about organizations, each is built on different assumptions about how organizations are structured and how they function. Research and development is unique in terms of its work activities, people, structure, and interfaces with other functional areas of the organization. The uncertainties that, on the one hand, are reduced by research but are, on the other, also evoked by it, create a form of organization that may provide indications for alternative structural approaches for organizations with uncertainties about technology. Understanding of the R&D organization provides a good set of assumptions for formulating the operating procedure for a manufacturing organization or a marketing organization.

In particular, R&D organizations can be characterized by uncertainty from technical areas both outside and within the larger organizations. Researchers such as Thompson (1967), Katz and Kahn (1966), and Weick (1969) have supported the notion that R&D organizations are open social systems that must deal with work-related uncertainty. Tushman (1976) suggests that in order for an R&D organization to deal with several sources of technical uncertainty, it must facilitate the gathering, the processing, and the exporting of information. In short, organizations must develop information-processing mechanisms capable of dealing with internal and external sources of work-related uncertainties.

R&D organizations can, in fact, be usefully seen as information-processing systems. Information processing can be seen as an ongoing problem-solving cycle involving each area of the R&D organization, the larger organization, and the external information world. Allen (1966) indicated that verbal technical communication is a particularly effective medium for the transfer of complex information and ideas, and that R&D organization can be seen as verbal communication networks. The idea of verbal technical communication is accentuated in R&D settings because technology is difficult to document and because of the information-gathering characteristics of a technologist (Allen, 1968).

Another characteristic of R&D organizations is that they are usually made up of groups or departments, which differentiate as the organization grows; that is, in order to realize economies of scale and benefits of specialization, subunits are usually created that have specialized tasks and/or deal with specific aspects of the organization's task environment (Katz and Kahn, 1966; Lawrence and Lorsch, 1967; Thompson, 1967). These subunits are usually interdependent to varying degrees and must share scarce resources, so their activities must be linked.

Efficient use of time is usually one of the most difficult areas to control in R&D organizations because the tasks of the subunits vary in their degrees of uncertainty. Galbraith (1973) suggests that this is so because tasks differ in their amount of predictability, and therefore in their amounts of information-processing requirements. For example, routine tasks can be preplanned and their information-processing requirements are minimal. Complex tasks that are not well understood or involve a number of unplanned exceptions cannot be preplanned and are therefore associated with greater uncertainty. Because the nature of the R&D organizational environment is dynamic and units often have little internal control, uncertainty is even greater. Also, because subunits in R&D organizations work together to coordinate research efforts, exchange ideas, and so on, the coordination and joint problem solving can be seen as another source of uncertainty. These uncertainties make it difficult to manage time effectively in R&D organizations. As tasks become less routine, the task environment becomes more dynamic, and as task interdependence becomes more complex, the subunits in R&D organizations are required to cope with both an increased amount of work-related uncertainty and increased information-processing requirements.

R&D organizations can be seen as differentiated, restricted communications networks. The differentiated areas usually have interdependent tasks, so communication between areas often occurs through technical roles that evolve to connect a project to other areas in the organization. The task characteristics that vary with the amount of information required and with the environment are not under the unit's control and are therefore potentially unstable.

6.2 MATRIX ORGANIZATIONAL MODEL

In the R&D organization, activities are being developed by research in a number of fields that lead to more direct and more intensive contacts with the other parts of the organization and with society. These contacts provide an organizational stimulus to forms of cooperation that display characteristics of a matrix organizational structure.

A prime characteristic of the matrix in an R&D organization is dual hierarchical responsibility. In particular, the responsibility of business management exists side by side with the responsibilities relating to the scientific starting point, objectives, and method of tackling the problem. Business management and scientific advisors represent their respective responsibilities both to the outside world and within the organization. For the first of these hierarchies the element of appointment will play a part, while for the second, selection leads to the recognition of a Primus interpares.

Alternatively, the scientific worker within an R&D organization can be seen as acting on the basis of belief in a system of concepts and assumptions that together form the matrix. Experiences are processed through this matrix, and sometimes it is adjusted. The scientists also have access to the matrices of others. For researchers, this system is not necessarily limited to one discipline. They make use of authorities or the consensus of opinions from other disciplines. The matrix therefore contains the paradigms they receive and their evaluations of sources of knowledge. The experiences that can be processed through this matrix may give rise to change in researchers' own opinions and evaluations of other opinions. The matrix concept is thus proposed for the R&D organizations based on the characteristics and relationships described.

The significant points concerning the matrix organizational model for R&D organizations are as follows:

1. In R&D organizations, cross-fertilization is regarded as useful for promoting the generation of ideas. A matrix organization helps promote personal effort and the independent operation of teams.
2. A matrix organization provides for direct and immediate work flow supervision on a product basis as well as specialization by function, which is often needed in R&D organizations.
3. A matrix organization permits management by results, communication among employees who have the greatest need to cooperate, and a single manager responsible for the entire research process. On the other hand, professionals and specialists can still have their home departments, the career satisfaction associated with working with peers, and the opportunity to work

for technical excellence. Authority is founded more on knowledge and skill than on formal organizational authorization.

4. Typical of R&D organizations are short-term projects in which functional specialists are assigned to complete the job in a specific length of time and adhering to fixed costs and performance standards. The matrix organization will help achieve expected output.
5. The matrix model can be used wherever innovative effort is needed to accomplish something that has not been tried before and that cannot be made routine.
6. Through the matrix model, the formulation of the objectives of research as a part of the total innovation process will reflect the dual appreciation for long-term scientific results and for application of short-term spin-offs.

6.3 FLEXED CONTINGENCY LEADERSHIP STYLE

In R&D organizations a hierarchy of objectives is necessary because of the complex relationships between groups. Differentiation and integration of tasks can then be carried out easily. In general, a manager in an R&D organization will be required to match a project's information-processing capacity with its information-processing requirements and specialize internal communication patterns to fit the information-processing requirements and the output of the task.

A flexed contingency leadership style with high consideration is therefore recommended for the R&D organization. A flex contingency leadership style requires a leader with the professional knowledge and skill to meet the requirements of the matrix organization model discussed in Section 6.2.

From the organizational point of view, three levels are recommended. The level at which by selection of means and objectives of the innovation are chosen, the level at which the conditions are created, the level at which one is responsible for the efficient and effective use of the means. The steering level is usually influenced by two environmental categories: the scientific world and the product/system groups. It is important therefore to understand these two categories when formulating objectives for R&D organizations. Short term spin-offs for application as well as long term results for scientific development have to be balanced. A flexed contingency style of leadership with high consideration will do best to satisfy this need. A third line in the program development comes from the grass root groups: who can appreciate the ideas that come from these groups, a task oriented leader with high consideration is needed to appreciate the ideas that come from these groups.

In R&D organizations, project task characteristics might affect the amount of communication within the project especially in situations where the locus of technical control is outside the project. Communication in the R&D organization, therefore, is indeed important that communication patterns within the organization are contingent on task characteristics. The task-oriented leader may be needed to manage communication specialized to fit the information requirements of the task. Patterns of technical communication may affect an organization's ability to attend to and deal with work related uncertainty. One important way of dealing with this is through verbal or oral communication because verbal communication is a particularly effective medium for the exchange of ideas, information, and concepts since it permits rapid feedback and recording of information. Verbal communication is therefore recommended for R&D organizations. Another technique for dealing with difficulties of communicating across organizational boundaries is the development of special technical roles. Information thus enters the organization through individuals occupying boundary roles who, in turn, channel this information with the organization. These special technical roles straddle several communication boundaries and represent an important network specialization to effectively link internal areas with external information domains.

6.4 DESIGNING A PRODUCT FOR AUTOMATION AND PRODUCTION

Product design for automation is an integrated procedure that requires the system and product designer to include manufacturing automation requirements into the overall design. The integrated procedure requires careful examination of all operations, parts, steps, and logistics of the product for automation capabilities and potential. The use of automation should always be to optimize cost, improve quality, and provide better manufacturing options and marketing.

6.4.1 Benefits of Product Design for Automation

When a product is designed correctly for automation, the following benefits can be obtained:

1. Reduction of production costs related to handling and storage, excessive inspections, rework and repairs, changeovers, and setups; reduction of the total number of operations that drive the assembly hours required.
2. Reduction of the total number of parts required for a product. Automation shortens the product introduction cycle and provides greater manufacturing flexibility.
3. Enhanced communications between all functional areas, such as research and development, manufacturing, and marketing.
4. Introduction of automation on the manufacturing floor. Product automation can also be done in stages.
5. Product design for automation allows the designer to justify and implement new technology based on a much better understanding of the product life cycle, the parts required, and the technology needed at each stage.

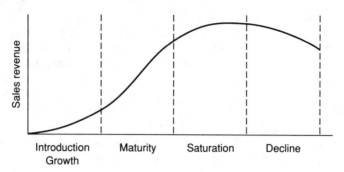

FIGURE 6.1 Product Life Cycle.

6.4.2 Product Life Cycle and Ideal Manufacturing Goals

In order to design a product for automation, it is important to understand the product life cycle shown in Figure 6.1. The product life cycle is a useful guide for assessing production requirements, the marketing plan, and introduction of new products, and for planning the firm's capacity and expansion requirements. The total length of the life cycle and the length of each of the stages can vary considerably from one product to another. Therefore, the designer may have to tailor the automation requirements to the various time periods.

Kuriloff and Hemphill (1978) describe the four distinct stages in the life cycle of a product or service as follows

- *Introduction/Growth Stage.* If a product or service is introduced and catches on, it may enter a period of rapid growth lasting several months or even years. Many firms may enter the industry during this period of expanding demand. Sales volume grows at an increasing rate, and profits for firms in the industry rise sharply. Competition also increases, but the entry of rival firms with their additional promotion efforts may actually enlarge the market.
- *Maturity Stage.* Eventually the level of market acceptance and sales volume reaches a peak. Sales revenue may continue to rise somewhat, but the rate of increase falls off, resulting in a decline in rate of profit. This is a stage of intense competition. Expenses for promotion become heavy, price cutting may occur, and consumers are pressured to be "brand loyal." Firms that cannot keep pace drop out of the market or are acquired by others. Large firms begin to dominate the industry.
- *Saturation Stage.* At this stage the consumers who can or will use the product or service are already buying—the market is saturated. Sales volume and profit in the industry begin to fall. Marginal firms have left the market, and the number of competitors stabilizes. The firm's promotion strategies concentrate on taking customers away from others rather than on enlarging the market.

- *Decline Stage*. Demand for the product or service falls at an increasing rate at the end of the cycle. Promotion is curtailed and becomes highly selective, and prices are cut to stimulate sales. More and more rivals drop out of the market. New products or services are developed to take the place of those that are now obsolescent.

The ideal manufacturing goals to consider when designing a product for automation should include but are not limited to the following:

1. Eliminate engineering changes for all systems designed for production.
2. Eliminate adjustments, screws, and cables. The screws requirements can be eliminated by using auto fasteners. Adjustments on parts or systems can be eliminated by designing and producing to clear defined tolerance.
3. Provide minimum accessories, modularity, self-test, and single power supply.
4. Provide assembly sequence that requires minimum direct assembly hours and zero defects on all products produced.

6.4.3 Factors Affecting Product Design for Automation

As shown in Figure 6.2, product design for automation can be influenced by internal and external demands on the design of parts. These factors need to be studied and understood on an ongoing basis to ensure that their negative impact on the designer's strategy is minimized. A teamwork approach among research and development, manufacturing, and marketing in addressing the impact of these factors on design strategy is also very helpful.

6.4.4 Minimizing Negative Impact from Product Design for Automation on Manufacturing Processes

In order to minimize any negative impact of product design for automation, the designer must adhere to the design rules presented in Chapter 3. In addition, the following issues must be recognized.

1. Product design for automation must not disrupt plant layouts, facilities, or other material handling systems.
2. There must be strong awareness of when the product determines the process and when the process determines the product. Designers must be aware that certain processes produce products not suitable for automation. The selection of a new technology, tool, or piece of equipment is just as important as the selection of the right process.
3. Efforts should be made to design processes and equipment with the product's properties and characteristics in mind.

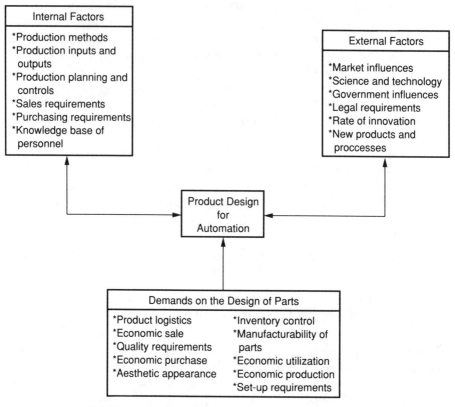

FIGURE 6.2 Factors affecting product design for automation

4. In situations where the manufacturing of a given product requires multiple vendors, effort must be made to cut down on cycle times, inspection requirements, double handling of parts, and movement of materials.
5. Product reliability, maintainability, serviceability, disassembly, appearance, obsolescence, and human factors must also be considered when automation requirements are being assessed.

6.4.5 Application of Value Analysis Technique in Designing a Product for Automation

The value analysis technique provides a useful alternative for examining and challenging the worth of every material, part, and operation involved in product design. When deciding on additional requirements for a feature or function, it is important to (1) clearly identify the need for the extra features, (2) provide a cost justification for the extra features, (3) perform a total evaluation of the worth and productivity of the additional features, and (4) explore different value alternatives to obtain the additional features. A good analysis of the desirability of additional features comprises seeking the right answers to the following questions:

- Does the additional feature or option contribute value?
- Is the cost of the additional feature proportionate to its usefulness?
- Does the feature need all of its proposed components?
- Is there anything better for the intended use of the new features?
- Can a usable part be made by a lower-cost process, method, or procedure?
- Can a standard product be found that will be usable?
- Is the right tooling, equipment, or machine used for the quantities needed?
- Is anyone purchasing the same features for less?
- Can another dependable supplier provide the feature for less?
- Are there costs other than those for material, reasonable labor, overhead, and profit?
- Does the quality of the additional feature or option contribute to value, or does it cause more rework and inspection?

It is also helpful to monitor value of parts and products through the four phases of the product life cycles specified in Figure 6.1. Actions are then implemented as necessary to control parts requirements by product life-cycle phases.

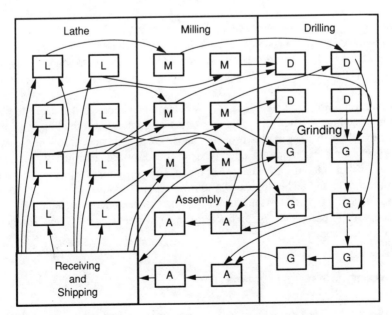

FIGURE 6.3 A typical layout pattern for machines.

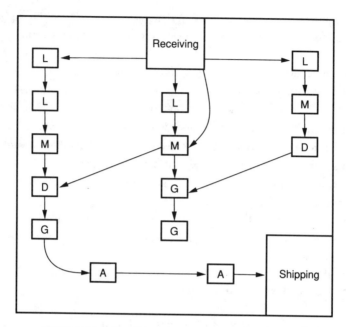

FIGURE 6.4 A typical GT layout pattern for machines

6.5 THE GROUP TECHNOLOGY CONCEPT

The group technology (GT) concept of productivity and overall efficient management of a production environment has surfaced in the last two decades. Ham (1976), Cho and Ham (1985), Rodriquez and Adaniya (1985), and Burbridge (1975), have discussed GT concepts and presented useful procedures for cell allocation and job scheduling.

Group technology is a technique for production management that seeks to obtain economic savings in batch and jobbing production similar to those already achieved using line flow in the simpler process industries and in mass production, while at the same time providing a better type of social system for industry, in which improved labor relations are easier to achieve. GT concepts permit the grouping of dissimilar tools and technologies in a unit or cell to produce an output. A typical process layout of machines and a GT layout are presented in Figure 6.3, 6.4, respectively. A case study of productivity measurement in a GT environment is presented in Chapter 8.

6.6 PRODUCT RELEASE METHODS

The phase-gate release method focuses on defining and addressing a specific product activity at a time. Activities are broken down by phases until each phase

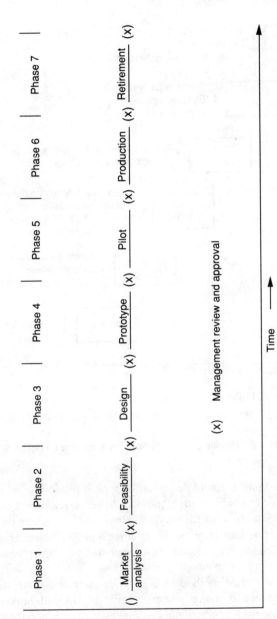

FIGURE 6.5 Phase-gate release method for new products.

116 MANAGING PRODUCT DESIGN AND DEVELOPMENT ACTIVITIES

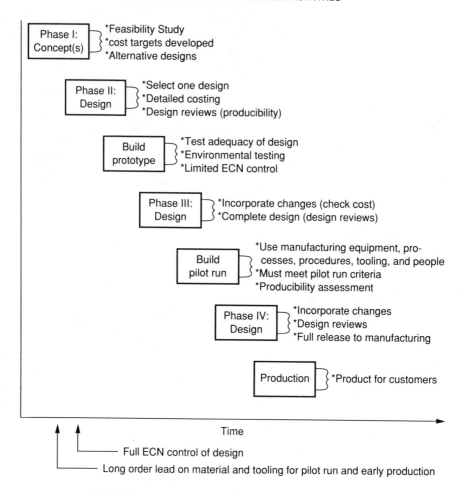

FIGURE 6.6 Parallel-phase method release for new products.

is completed the next phase cannot be started. A typical phase-gate release method is presented in Figure 6.5. This method provides lower development costs through detailed review at each phase of the product release cycle, reduces design errors, provides control, and defines responsibilities at each phase. However, the product release cycle takes longer, and the phase-gate method limits new product introductions and manufacturing involvement.

The parallel-phase release method focuses on defining and addressing multiple product activities in one or more phases. Activities are still broken down by phases, but a dual approach to completing required activities is permitted. A typical parallel-phase release method is specified in Figure 6.6. Implementation of this method often speeds new product introductions, improves communication, provides early manufacturing involvement, and allows for more product introductions. The disadvantages associated with this method are a potential for greater risk and

6.7 EARLY MANUFACTURING INVOLVEMENT (EMI) IN THE TECHNOLOGY DESIGN PHASE

increased costs, a less structured environment, and the requirement that units of organization share the burden of dual responsibilities.

A careful analysis of the manufacturing and marketing problems associated with new technologies reveals that close communication is needed between R&D, manufacturing, and marketing organizations. It is typical for an R&D organization to generate a series of engineering changes on a new product without realizing the total impact on the manufacturing processes and the lead time required to satisfy customer orders. An understanding of the R&D organization, which creates ideas, provides a good set of assumptions for formulating the operating procedure for a manufacturing organization, which produces, and a marketing organization, which sells. EMI provides a viable medium through which R&D, manufacturing, and marketing can deal with internal and external sources of work-related uncertainties involved in the design of a new technology.

6.7.1 EMI Defined

EMI can be defined as a concept that requires the total involvement of research and development, manufacturing, marketing, and other functional areas in all phases of

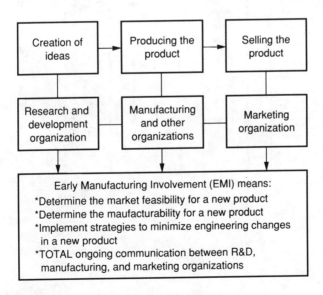

FIGURE 6.7 Components of EMI in technology product design.

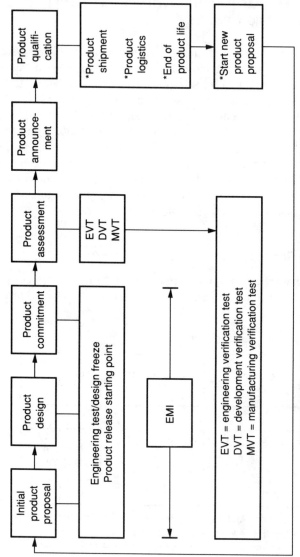

FIGURE 6.8 EMI coordination path in technology product design.

EARLY MANUFACTURING INVOLVMENT (EMI) IN THE TECHNOLOGY DESIGN PHASE 119

a new technology product design. The EMI concept is shown schematically in Figure 6.7. The EMI coordination path in a typical technology product design, test, and release cycle is shown in Figure 6.8. The EMI effort is needed from initial product design through the engineering test cycle.

6.7.2 EMI Objectives

The primary purpose of EMI is to influence the design of a new technology product for manufacturability, reliability, and quality. In addition, EMI ensures that a detailed release plan is available for new products, strives to minimize engineering changes, maximizes use of common parts, maximizes compatibility between new and old manufacturing processes, and provides adequate product management systems. The components of EMI objectives are presented in Figure 6.9.

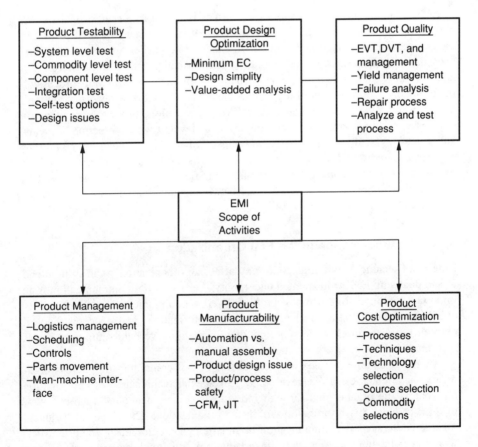

FIGURE 6.9 Some components of EMI objectives.

6.7.3 Matrix Organizational Model for EMI

EMI activities require direct and intensive contacts with each subunit of the total organization. The contacts provide an organizational stimulus to forms of cooperation that display characteristics of a matrix organizational structure. A prime characteristic of the matrix in EMI organization is dual hierarchical responsibility. The EMI analyst has responsibility for researchers, production personnel, and sales representatives in regard to their effects on the organization, side by side with the responsibility relating to the scientific problems.

The implementation of EMI activities requires support from all functional areas. The matrix organizational concept allows a functional specialist to report directly to a program or product manager. This facilitates quick resolution of problems that would otherwise take several months to resolve using traditional hierarchical organizations.

6.7.4 Leadership Style for EMI Projects

In EMI organizations, a hierarchy of objectives is necessary because of the complexity relationships between groups. Differentiation and integration of tasks can then be carried out in a simple manner. In general, a manager responsible for EMI activities will be required to match a project's information processing requirements and specialize internal communication patterns to fit the information processing requirements and the output of the task. A flexed contingency leadership style with high consideration is recommended for EMI organization. From the organization point of view, three levels are recommended: the level at which selection of means and objectives of the EMI activities are chosen, the level at which the conditions are created, and the level at which one is responsible for the efficient and effective use of environmental categories: the scientific world and the product/system groups. It is important to understand these two categories when formulating EMI objectives.

6.7.5 Communication in the EMI Organization

Patterns of technical communication may affect an organization's ability to attend to and deal with EMI work-related uncertainty. Verbal or oral communication is an important tool in dealing with this problem. Verbal communication is a particularly effective medium for the exchange of ideas, information, and concepts since it permits rapid feedback and recording of information. Verbal communication is recommended for EMI organizations.

Another technique for dealing with difficulties of communicating across organizational boundaries is the development of special technical roles. Information thus enters the organization through individuals occupying boundary roles who, in turn, channel this information within the organization. These special technical roles straddle several communication boundaries and represent an important network specialization of effectively linked internal areas with external information domains.

SUMMARY

Research and development organizations can adapt better to changing conditions if they adopt an open communication structure or network. Innovation is generated primarily by cross-fertilization of ideas among the various fields of disciplines, which allows individuals to develop the ideas of others, resulting in the discovery of unknown areas. Through an open system, researchers have the opportunity to learn about the development in other environments and different disciplines. In contrast to pure research in one discipline, where centralized organizations have arisen in an attempt to promote effectiveness, the fluid organization within which the various frames of reference come together has proven to be more effective for innovation. A closer communication between R&D, manufacturing, and marketing is also needed to reduce the cost of new products to consumers. Closer integration of design, development, and manufacturing activities is highly recommended. EMI efforts address such integration efforts.

Several inhibitors can impact the design efforts for automation. These inhibitors include disparity in tools; geographical location; lack of experience in design, manufacturing, and controls; lack of early manufacturing involvement in the design and development process; lack of teamwork among functional organizations; and lack of management support for new ideas. The design for automation of products can easily be implemented by facilitating processes, procedures, and systems that emphasize total manufacturability for product, by providing adequate means for test and tool compatibility, by encouraging more vertical organization structure, by adopting specific design ground rules, and by educating the total organization on automation concepts and techniques. Designing a product for automation is made possible by a top-down commitment within an organization on all key requirements for the new product.

QUESTIONS AND PROBLEMS

6-1 What characteristics make research and development differ from other types of organizations?

6-2 Assume you are the plant manager at a factory producing personal computers to customer orders. How would you implement the concept of early manufacturing involvement within your organization?

6-3 Develop an organization model for R&D activities.

6-4 Suggest a leadership style for research and development activities. How would you implement the suggested leadership style in a large corporation, in a small firm, and at the department level?

6-5 What types of activities should an EMI group focus on?

6-6 What leadership traits should be sought in the selection of a project manager responsible for EMI activities?

MANAGING PRODUCT DESIGN AND DEVELOPMENT ACTIVITIES

6-7 Recommend a strategy for minimizing the total number of engineering changes in technology design.

6-8 How can the communication channels between research and development, manufacturing, and marketing be enhanced?

6-9 How would you proceed to set up an organization on a matrix model?

6-10 What are the common problems involved in managing R&D activities?

6-11 Explain the concept of product design for automation.

6-12 Assume you have been given the role of a design engineer in a computer firm. How would you proceed to design a new personal computer for automation?

6-13 Explain the product life cycle phases. Should a system designer incorporate the cycle limitations in a new product designed for automation?

6-14 Define the concept of group technology and discuss its limitations in a production environment.

6-15 Suggest a procedure for allocating parts and machinery in a group technology environment.

6-16 Discuss the various factors affecting product design for automation and suggest a solution strategy for minimizing the impact of these factors.

6-17 Define and discuss the application of value analysis technique in product design issues.

6-18 Explain how you will proceed in applying the value analysis technique in the design and implementation of a 7535 Robotic device.

6-19 What are the roles of top and middle managers in designing a product for automation?

6-20 Suggest a procedure for assessing a product designed for automation.

7
MANAGING CURRENT-GENERATION TECHNOLOGY

This chapter presents ideas for managing current-generation technology at the enterprise level. Some of the most commonly used technology, such as expert systems, computer- aided design, computer-aided manufacturing and robotics, is discussed.

7.1 KNOWLEDGE-BASED EXPERT SYSTEMS

Both the manufacturing and service work environments have always depended heavily on the use of human expertise to produce goods and services. In order to maintain that human expertise, considerable efforts are required in training of the workforce. Moreover, the human expertise is hard to transfer, hard to document, and can at times be unpredictable. Knowledge-based expert systems (KBES), developed in the field of artificial intelligence (AI), have strong potential (and some proven capability) for reducing training cost, maintaining consistent expert knowledge, and improving productivity and the quality of the task performed. This section focuses on the presentation of key design rules that will enhance the effectiveness and use of knowledge-based expert systems.

7.1.1 KBES Definition, Structure, and Configuration

A knowledge-based expert system is a representation of the expert's heuristic reasoning through computer programs that emulate the behavior of humans in solving problems normally thought to require experts for their resolution. The system structure consists of six parts.

1. The *knowledge base* is composed of formal rules, definitions, facts, and expert-level heuristics for solving or diagnosing problems. The knowledge involved represents the expert's good judgment, rules-of-thumb, and educated guesses based on experience. The knowledge base is a software program that contains facts and rules that the knowledge engineer has gleaned from experts in the field.
2. The *inference engine* provides an organized procedure and controls for using the knowledge base to solve problems. It contains the acquired knowledge and strategies to solve problems, and provides an interface with other operating systems.
3. The *working memory* consists of relevant data that describe the particular problem being addressed. In some KBES applications, the working memory can be viewed as a part of the knowledge base.
4. The *KBES user* is the person who interfaces with the expert system to determine the capabilities and parameters of the system cell.
5. The *domain experts* are the resources for the knowledge of the system; they use their expert knowledge to develop the system.
6. The *knowledge engineer* is the person who synthesizes the facts and heuristics of the domain experts, so that the knowledge of the experts can be captured and computerized.

The basic structure of an expert system is presented schematically in Figure 7.1.

7.1.2 KBES Impact on Productivity

Without the human mind, the successful development and use of the various technologies will not happen in the first place. Human minds do some things extraordinary well; the unique human quality of intelligence extends beyond our ability to reason, judge, question, and make decisions. But maintaining human expertise requires considerable effort in training the workforce. In addition, human expertise can be unpredictable. Knowledge-based expert systems should be viewed as another mechanism to expand human physical and intellectual capabilities. KBES provides useful benefits in the following situations:

1. Where expertise is not available on a reliable and continuing basis.
2. Where experts are difficult to train.
3. Where expert knowledge is very expensive.
4. Where expert knowledge is in high demand.
5. Where there is no significant dynamic variation in production or service process.
6. Where there is a high risk of losing sensitive, expensive, and useful information based on experience.

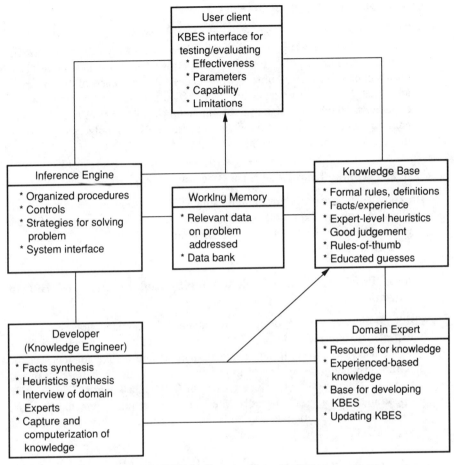

FIGURE 7.1 Expert system structure.

However, the successful application of KBES depends on a sound method for representing knowledge in a symbolic fashion in a computer-like system. Most of the current KBES's have been developed using key concepts to mimic a human's logical reasoning process. Fergenbaum and McCorduck (1983) describe one such technique as a set of *if-then rules*. An if-then rule is a form of antecedent-consequent logic, in which an outcome (i.e., consequence) becomes the antecedent of the next decision rule. Other rules for KBES design can be found in the works of Davis (1982), Rich (1983), and Fichtelman (1985).

7.1.3 Design Rules

In order to design knowledge-based expert systems that are effective and efficient, the following design rules are suggested as a guide for system designers.

Rule 1: Obtain the Right Knowledge Base

A KBES knowledge base must be based on accurate information from history, experience, and expert judgment that has been adequately tested and demonstrated successful results. For simple repetitive tasks, the knowledge base on a sample size of 30 humans with successful experience of five years is adequate. Nonrepetitive tasks are very difficult to handle; several years of experience, thousands of iterations (including data gathering), and experiences of several hundred humans are needed. KBES is only as effective as the knowledge base used in the design.

Rule 2: Form the Knowledge Base Procedure

The knowledge base procedure should include the consistent and correct methodology for solving the problem being addressed. The designer of KBES must ensure that the procedure is based on streamlined useful information obtained from Rule 1.

Rule 3: Provide Adequate Structure for Systems Prompts and Human Socialization

KBES prompts should be designed to emulate the human phrases they are expected to represent. Prompts that are not clear pose difficult problems to the users of KBES. It can create limitations and dislike for the system. KBES designers must endeavor to use program statements that are clear and concise. The package of programs used in KBES must also provide a wide opportunity for a user's friendly interface. This requires selected prompts to be interesting, rational, and believable, and to cover a wide domain of application.

Rule 4: Provide Timely KBES Responses

KBES should be geared to increase productivity. Human waiting time for the system must be kept to a minimum if the system is going to be productive. Unnecessary iterative procedures, rules, and do-loops in programming should be avoided. System response time and sharing options should be evaluated in several prototype programs through the use of the value analysis technique.

Rule 5: Provide Adequate Explanation and Documentation for All System Variables

System designers should provide users the ability to query and interface in a friendly manner. This requires a knowledge base that has a help center with a capability to guide and answer several possible options from a user. The documentation process involved in the use of KBES can be made simple through the use of illustrative flow charts and graphic displays and symbols.

Rule 6: Provide Adequate Time-sharing Options

In order to increase productivity of a KBES, the system must have adequate provision for simultaneous use by several workers. The optimum measure of KBES time-sharing options can be expressed as:

Tac_{ijt} = Total available KBES time-sharing options of KBES i at site j in period t.

Nf_{ijt} = Number of firms or workers sharing KBES i at site j in period t.

TSR_{ijt} = Time-sharing ratio of KBES i at site j in period t.

$$\text{TSR}_{ijt} = \frac{\text{Tac}_{ijt}}{\text{Nf}_{ijt}} = 1$$

Optimum TSR_{ijt} is obtained through examination of several iterations with possible options for sharing KBES resources.

Rule 7: Provide Adequate User Interface

KBES should allow workers the opportunity to learn new skills or enhance the existing knowledge base. The control options provided on the system must be free from psychological and stress problems. The KBES work station must be ergonomically feasible for workers' use.

Rule 8: Provide Intersystem Commmunication Ability

The effectiveness of KBES is greatly enhanced if its various systems can communicate with each other under one system network. The ability of the various KBES's to understand the same natural language is of equal importance.

Rule 9: Provide Automatic Programming Ability and Controls

Automatic programming ability and control will minimize system downtime and increase KBES efficiency. The various controls should also have the capability to advise, alert, and warn users of potential events.

Rule 10: Provide Flexibility for Ongoing Maintenance and Update

New techniques, approaches, and methodologies for solving problems always arise periodically. KBES design must allow for flexibility in updating the experience base of the system and procedure and system options.

7.1.4 Managing a KBES Project

The KBES project management phases shown in Figure 7.2 are recommended as a guide to potential users of KBES technology. KBES application experts, knowledge

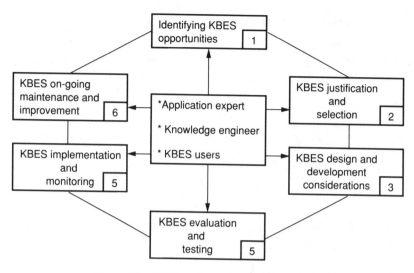

Figure 7.2 KBES project management phases.

engineers and users must work together to address the issues involved in all six phases. It is equally important to understand the limitations of KBES as well as its advantages in the work environment.

Phase One: Identifying KBES Opportunities. At this phase, the KBES designer and user review the current mode of business processes and operations and identify appropriate operational units for KBES applications:

1. Review areas in your enterprise where large units of expertise, time, capital, or other resources are used and determine methods by which usage might be reduced.
2. Determine if the expertise associated with the areas mentioned in Step 1 can be reduced to procedures from which rules may be identified.
3. Analyze the rules and premises to determine if associated actions could be defined.
4. Make a flow chart of the decision-making process of the first three steps. Identify variances or problems that might be reduced by using an expert system.
5. Cost out the process with and without an expert system. Prioritize candidate applications.

The KBES opportunities identified should be classified into two areas: repetitive and nonrepetitive tasks. The opportunities identified should be prioritized with consideration to KBES design complexity, anticipated costs, and potential benefits expected from the KBES technology.

Phase Two: KBES Justification and Selection. KBES justification should be done before and after the design of the system. Before the full KBES operation system is designed, an attempt should be made to determine in quantitative and qualitative terms the value of potential benefits to be derived from a specific KBES application. A rough estimate of design costs and resources required is essential to avoiding overpriced KBES technology. The justification of later stages can be performed using economic analysis, ratio of multiple users (cost and benefits for KBES sharing options), and other intangible benefits (problem-solving capability, training tool, security of knowledge, and information base reliability). The following KBES justification model is recommended.

K_{it} = Payback period required for a specific KBES i, utilized for a specific task t

TI_{it} = Total investment (development + testing cost + installation and training cost + system accessories cost + KBES price) in KBES i for task t

R_{it} = Labor and material costs replaced by KBES i for task t

M_{it} = Maintenance cost involved in KBES i, for application in task t

DC_{it} = Annual depreciation cost of replaced system/equipment through the application of KBES i for task t

DP_{it} = Difference in productivity level (KBES application versus non-KBES method) resulting from the application of KBES i for task t

$K_{it} = TI_{it}/[(R_{it} - M_{it}) + DP_{it}(R_{it} + DC_{it})]$

Managerial judgment should be used to determine the right level of intangible benefits to be derived from KBES applications.

Phase Three: KBES Design and Development Considerations. The ten design rules already discussed in this section are recommended for effective and efficient design of KBES technology. In addition, KBES development should be based on formulated criteria and processes. The criteria to consider should include but not be limited to the following:

- Thorough understanding of application environment.
- Task complexity.
- Availability of domain expert.
- Degree of task cognition.
- Expected benefits from KBES.
- KBES flexibility and ease of implementation.

The development process requires that the opportunities identified in phase one be organized in the form of English-like sentences and procedures. Based on a process of conceptualizing knowledge, the KBES rules are then formulated and a prototype is developed.

Phase Four: KBES Evaluation and Testing. At this phase, the KBES rules and procedures are evaluated and tested. The key task at this phase is to verify the accuracy of the knowledge-based rules using dedicated domain experts. The KBES rules are then revised based on feedback results. The evaluation and testing process requires interaction among the knowledge engineer, domain expert, and the end user in redefining the appropriate rules to provide efficient and effective KBES technology.

Phase Five: KBES Implementation and Monitoring. Successful implementation of a KBES prototype or final system requires the total involvement of the user community. This can be accomplished through a formal meeting between the knowledge engineer, domain experts, and users to clarify KBES goals and objectives, application potentials, resources required, process impact assessment, and benefits to be derived. Training of the workforce on how to use the new KBES technology is essential. Training emphasis should be on:

- Interpretation rules.
- How to analyze a particular problem using the system.
- KBES–user interaction procedures.
- How to execute KBES problem-solving logical routines or steps.
- Help procedure.
- Maintenance.
- Security requirements.

Phase Six: KBES Ongoing Maintenance and Monitoring. At this phase, ongoing assessments of KBES rules, procedures, prompts, response time, documentation, sharing options, and controls are performed. Changes in the application environment are documented and used to modify the KBES rules. Maintenance and monitoring require attention to the following details:

- Changes in task execution processes.
- Changes in troubleshooting techniques.
- Changes in skill level of users and domain experts.
- Changes in multiple-user ratio and requirements.
- Changes in initial problem definition.
- Changes in solution space.

The KBES maintenance process is a continuous one that interrelates actions in all six phases to improve the design of a new system or the modification of an existing knowledge base.

7.1.5 KBES Application Opportunities

KBES technology will play a leading role in managing the information-based factory of the future. Potential application opportunities include but are not limited to the following areas:

- Defect isolation tasks
- Failure analysis tasks
- Financial advising
- Production scheduling
- Operator training
- Advisor for measuring process variables
- System control and shutdown
- Diagnosis of machines
- Process control advisor
- Error cause removal advisor
- Solution space mapping
- Simulation advisor
- Patient care advisor
- Capital budgeting
- LSI and VLSI layout
- Productivity evaluation/analysis
- Client controls
- Circuit analysis
- Maintenance activities
- Drilling advisor
- Process setup advisor
- Host-to-host system mapping
- Data control
- Line balancing advisor

7.2 ROBOTICS

Industrial robots with point-to-point controls for simple material handling tasks were first introduced commercially in 1959, and the first robot with path control capability appeared in 1961 (Ayres, 1988). These robots were suitable for a number of purposes, including spray painting, spot welding, arc welding, and investment casting.

Like those early machines, today's typical industrial robot consists of three basic components.

132 MANAGING CURRENT-GENERATION TECHNOLOGY

Power Supply: Provides and controls the power for the actuators of the manipulator.

Control System: Provides the sequencing and coordination of the various axes of the robot. It also serves as a communication source and linkage with external systems (machines and objects).

Manipulator: Uses the axes of motion to perform the actual work of the robot in reaching, grasping, moving, placing, and removing parts or completing a set of tasks within an operation.

Sumanth (1984) lists the following advantages of using robots in industrial settings:

1. Robots can work three shifts a day, uninterrupted, without being paid overtime.
2. Robots can perform tasks that are arduous if not impossible for a human being (examples: handling heavy parts or hot metal slabs, and working in poisonous, radioactive, and other hazardous chemical atmospheres).
3. Robots can improve quality by providing consistency in manufacturing operations.
4. Robots can save substantially on wages, salaries, and fringe benefits.
5. Robots can help in overcoming the shortage of blue-collar workers when a society produces increasing numbers of college graduates, none of whom wish to work on shop floors.
6. Robots can help integrate CAD/CAM, automated guided vehicle systems, and flexible manufacturing systems to pave they way for the ultimate development of totally integrated, automated manufacturing facilities.

Suggested applications of robots in industry are:

- Spot Welding
- Spray painting
- Arc welding
- Grinding/deburring
- Material handling (loading and conveyance)
- Machine loading/unloading
- Assembly (e.g., assembling steering locks)
- Inspection (checking container labels, inspecting parts, gauging)
- Fish farming (feed and monitor fish)
- Firefighting (where heat and smoke would prevent the development of firefighters with conventional equipment)
- Die casting
- Filling/sealing/packaging/gluing/gasketing

- Warehousing
- Cake making and decorating
- Nuclear reactor operation
- Restaurant operations (e.g., making sushi)
- Casting (pouring hot metal into molds; inserting parts into furnaces)
- Cutting/machining
- Polishing/surface preparation

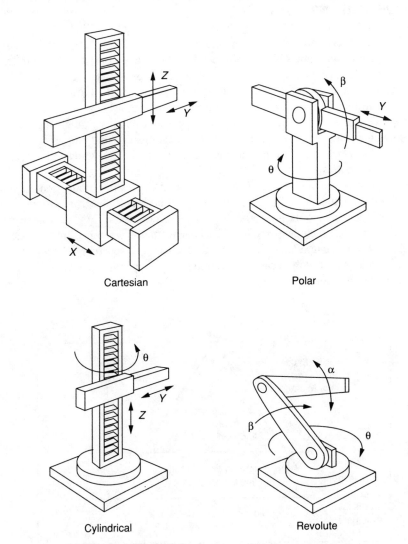

FIGURE 7.3 Robot classification by configuration and geometry

Robots can be classified by function or by configuration. Classification by function is usually done by programming path, such as servo controlled, continuous path, point-to-point, non-servo, and mechanical devices. The classifications by configuration are, cartesian, polar, cylindrical, and revolute, all of which are shown in Figure 7.3. A step-by-step approach for implementing a robotic device in the work environment is presented in one of the case studies provided in Chapter 8.

7.3 COMPUTER-AIDED DESIGN (CAD) AND COMPUTER-AIDED MANUFACTURING (CAM)

CAD is an integrated computer system that helps system designers to design processes, products, and objects. A typical CAD system consists of a mainframe computer, CRT, keyboard, plotter pen, pointer, and pen plotter. Compared to the manual method of designing objects, processes, and products, CAD offers a better approach in that it minimizes chances of errors in design, helps in the analysis of multiple design alternatives, and provides a means for obtaining quick and efficient design. Chorafas (1987) discusses CAD classifications in detail and points out that CAD performance will continue to increase, especially CAD adapted for 16-bit personal computers.

CAM can be viewed as high-level system that carries out production planning and scheduling functions for a manufacturing plant and generates programs for individual machine tools and other cells. In recent years CAM has been used for activities that include parts logistics management, assembly line balancing, machine loading and sequencing, operations control, inspection and defect mapping, inventory control, and capacity planning. CAM operates by manipulating data fed through the mainframe from control cells or machine centers. Each computer

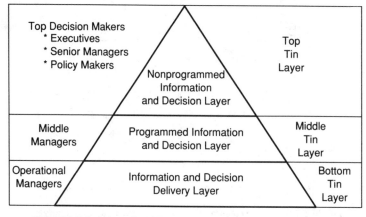

FIGURE 7.4 Triangular information network of operations (TINOP).

program in CAM is designed to support a certain application, such as inventory control and defect analysis. When implemented properly, CAM helps organizations improve productivity and quality and reduce inventory costs, turnaround time for processing data, and idle time through better scheduling of machines, operations, and other activities. In order to implement CAM properly it is recommended the following be available: adequate hardware, appropriate software, technical knowledge of computer programs, training methods for potential users, and adequate maintenance facilities.

7.4 COMPUTER-INTEGRATED MANUFACTURING (CIM)

The concept of CIM has been evolving as firms seek a complete solution to the management of the order entry, manufacturing, and invoicing processes. CIM can be viewed as a fully integrated CAD/CAM that provides the end user with the ability to perform several functions, including process automation, drafting, order entry, manufacturing process planning, production scheduling and control, inventory management, invoicing, and assembly line balancing. A complete network of CIM should have a complete data interface between each manufacturing cell, product design, manufacturing process planning, production operations management, shipping, marketing and invoicing control, parts logistics, and vendor requirements. Implementation steps for CIM and its advantages are discussed in details in the case studies presented in Chapter 8.

7.5 INFORMATION-BASED ORGANIZATION STRUCTURES

As the increasing use of computer technology in the workplace creates an information-based network of operations (IBNOP), the tall pyramid structure of organizations will flatten. Organization settings will become technology driven, and hierarchies will be streamlined to three layers in order to respond to the dynamic needs of the ever-changing technologies. The new organization structure shown in Figure 7.4 will be characterized as the triangular information network of operations (TINOP). TINOP will comprise three layers of technology information network (TINs).

7.5.1 TINOP Layer One: Nonprogrammed Information and Decision Layer

The top layer of TINOP will be responsible for nonprogrammed information and decision-making processes. There will be no cut-and-dried procedure for handling information at this level because the problems involved have not arisen before and because they will be very complex. Management at this level will involve the analysis of system design options; it will also require managers to provide goals and objectives and tools to monitor the total system and layer performance.

Effective management of the requirements of this layer will require (1) training of decision makers on heuristic problem-solving techniques, (2) the use of heuristic computer programs, and (3) training on the traditional methods of creative judgment, operations management, trade-off analysis, and strategic analysis of system performance and behavior.

7.5.2 TINOP Layer Two: Programmed Information and Decision Layer

The middle layer of TINOP will be responsible for programmed information and decision-making processes and procedures that govern the daily smooth running of the total system operations. Programmed information and decisions are repetitive in nature. The routine information and decision problems that occur in this layer lend themselves to a formalized procedure that will be available to handle such problems. Effective management of the requirements of this layer will require training of decision makers on information simplification, quantitative techniques for handling high volumes of information, operations improvement, and logistics management. An example of a computer-aided decision process enhancement model that is recommended for use in a high-technology environment is presented in Figure 7.5.

7.5.3 TINOP Layer Three: Information and Decision Delivery Layer

The bottom layer of TINOP is responsible for implementation of all information and decision processes. The delivery task involves carrying out the orders of the top and middle layers as well as ensuring that the final output of the system or organization gets delivered to the customer. Effective management of the requirements of this layer will require training of decision makers on control procedure and techniques for large-scale information project management, information network improvement techniques, input-output management in high-technology environments, and operations management techniques.

7.5.4 Handling TINOP Challenges

There is no doubt that decision makers of work organizations face tougher challenges in an information-based environment. The triangular information network of operations can be handled if there is an on-going system management approach in place for training the workforce, carefully and appropriately selecting decision makers, and managing the dynamic changes in the technology life cycle. Simon (1960), Webber (1975) and Drucker (1974) provide useful techniques for managing tasks, responsibilities, and systems. Useful quantitative techniques for decision analysis can be found in the works of Morris (1977) and Beer (1966). Techniques for selecting, justifying, and maintaining cost-effective high-technology-based systems are discussed in Chapters 4 and 8. Dooley and Stout (1971) describe seven

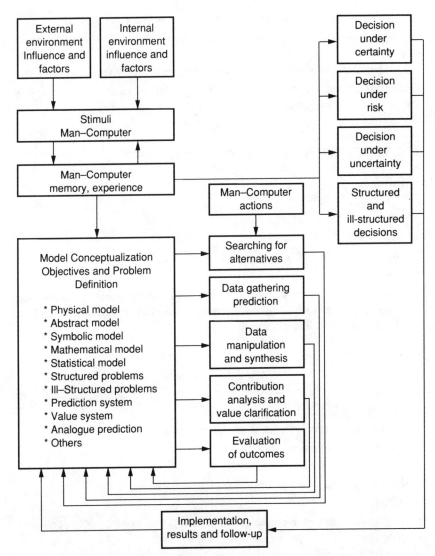

FIGURE 7.5 Computer-Aided decision enhancement model.

important points about successful computer-based automation (CBA) in manufacturing:

1. Success in CBA requires complex time planning and large time inputs.
2. CBA should be started at a simple level.
3. Care must be taken not to be locked into preconceptions about problems.

4. Demands for accuracy should not be excessive in the early stages.
5. A project team combining technical, managerial, and leadership qualities is required.
6. CBA should increase knowledge about production processes.
7. Deficits in memory capacities after use begins is the most common failure and should be guarded against.

7.6 AREAS OF RAPIDLY CHANGING TECHNOLOGY†

Components

1. Semiconductors
   Technology push:
   a. Integrated chips: VLSI design and fabrication, VHSIC
   Market pull:
   a. Custom chip design applications in electronics, computers, consumer appliances, industrial control, communications, aircraft and missiles, automobiles and transportation

Devices

2. Computers
   Technology push:
   a. Supercomputer architecture: parallel processing
   b. Computer peripherals: printing, memory, display
   c. Computer graphics and three-dimensional display
   d. Expert systems, software, and user friendliness
   Market pull:
   a. Segmented computer markets: mainframes, minicomputers, microcomputers
   b. Applications markets: business and office systems, manufacturing systems, scientific systems, personal computers, home entertainment and information
3. Robots
   Technology push:
   a. Manipulation and control
   b. Sensing: vision, tactile

---

†This section is adapted from F. Betz, Managing Technology, © 1987, pp. 97–100. Reprinted by permission of Prentice-Hall, Inc., Englewood Cliffs, NJ.

c. Flexible manufacturing: tools, materials, scheduling, handling
   d. Production and sales system integration

   Market pull:
   a. Automobile
   b. Aerospace and defense
   c. Electronics

4. Lasers

   Technology push:
   a. Lasing techniques and materials: frequency and power
   b. Laser tools

   Market pull:
   a. Laser communications: transmission, fiber optics
   b. Optical logic devices and circuitry
   c. Holographic imaging and measurement
   d. Laser tools
   e. Laser weapons

5. Scientific instrumentation

   Technology push:
   a. Nuclear magnetic radiation (NMR) measurement and imaging
   b. Synchrontron radiation
   c. Millimeter, infrared, and ultraviolet radiation sensing and measurement
   d. Automated instrumentation
   e. Remote sensing
   f. Computerized databanks and shared models
   g. Automated testing

   Market pull:
   a. University research
   b. Aerospace and defense research
   c. Chemical and petroleum industries
   d. Medical and pharmaceutical industries
   e. Electronic and computer industries

Processes

6. Optical circuitry

   Technology push:
   a. Fiber-optic transmission
   b. Laser equivalents of transistors: bitable devices, photronics

   Market pull:

a. Communications
b. Defense
7. Genetic engineering and biotechnology

   Technology push:
   a. Recombinant DNA techniques
   b. Tissue culture and monoclonal antibodies
   c. Genetics
   d. Fermentation and separation

   Market push:
   a. Scientific instrumentation and products
   b. Medical products: analysis and treatment
   c. Agricultural products and processes: plant stocks, pesticides, animal breeding, vaccines
   d. Specialty chemicals

8. Chemical processes

   Technology push:
   a. Instrumentation and chemical analysis
   b. Molecular design of compounds and reactions
   c. Laser-induced chemical reactions
   d. Catalysts, separations, multiphase processes
   e. Coal-and-biomass-feedstock chemical processes

   Market pull:
   a. American chemical industry
   b. Japanese and European chemical industries
   c. Third-world chemical industries
   d. OPEC

Systems

9. Communications

   Technology push:
   a. Distributed processing: local and global computer networks
   b. Computer security
   c. Integrated communications: voice, video, data

   Market pull:
   a. Business: automated office
   b. Telephone and data transmission
   c. Media: television and movies
   d. Publications: technical and general and video text

e. Databases: scientific and consumer
10. Computer-integrated manufacturing
    Technology push:
    a. Color computer graphics
    b. CAD design principles and algorithms
    c. CAM and flexible manufacturing tools
    d. CAD/CAM peripheral equipment
    Market pull:
    a. Aerospace and defense
    b. Electronics
    c. Automobiles
    d. General manufacturing
11. Military systems
    Technology push:
    a. VLSI and VHSIC electronics
    b. Laser weapons
    c. Composite materials
    Market pull:
    a. Smart weapons
    b. Electronic warfare
    c. Star Wars
    d. High-performance materials

Resources

12. Materials
    Technology push:
    a. Organic and metallic composites
    b. Ceramics: electronics, high-temperature
    c. Polymers: materials, fibers, adhesives
    d. Conducting plastics
    e. Semiconducting materials and processes
    f. Low-temperature materials and processes
    Market pull:
    a. Textiles, fabrication, construction
    b. Aerospace and defense
    c. Automobiles
    d. Electronics, computers, and communications
    e. Scientific instruments

LOOKING AHEAD

More revolutionary technology is expected to have both positive and negative consequences on productivity, quality of working life, and society's values in general. The organization of the future can be described as comprising horizontal and vertical information networks driven by computers, expert systems, and other high technology with minimum human intervention. The organizational operating environment will be characterized by high capital investment in technology, knowledge professionals and managers, and skilled domain system experts and operators. Among all modern technologies, knowledge-based expert systems and computer-integrated manufacturing (CIM) will have the most significant impact on the organization of the future. They will constitute the leading edge of technology required to manage the production of goods and services. The need for preservation, dissemination, and application of knowledge is the reason for the expected fast growth of the expert systems technology. Other technologies such as CIM, laser, and optical devices are also expected to play a vital role.

QUESTIONS AND PROBLEMS

7-1 Define knowledge-based expert systems.

7-2 Identify 10 business areas where knowledge-based expert systems can be applied.

7-3 What are some of the limitations of knowledge-based expert systems?

7-4 Develop an implementation strategy for a knowledge-based system designed for (a) an insurance task, (b) a production task, and (c) a bank clerical task.

7-5 How would you implement the knowledge-based design rules in a small manufacturing plant that wishes to use such a system to reduce training costs?

7-6 What are some of the implementation problems involved in knowledge-based expert system applications?

7-7 Outline the major advantages and disadvantages of using CAD/CAM, robotics, CIM, and expert systems.

7-8 Develop a set of subject criteria for evaluating CIM systems.

7-9 What are the implications of "total automation" based on centralized networks of information systems?

7-10 Suggest procedures for balancing the organizational requirements that are driven by new technologies such as robotics and expert systems.

7-11 Discuss the dependencies involved in phasing in new technologies and phasing out existing ones.

7-12 Discuss the role of professional managers in the information-based factory of the future.

8

CASE STUDIES IN TECHNOLOGY MANAGEMENT

This chapter presents case studies that deal with problems of justifying and implementing new technologies. Ergonomics and quality of working life issues in technology management are also discussed in some of the case studies presented.

8.1 CASE STUDY ONE: PROBLEMS OF TECHNOLOGY LEADERSHIP—XEROX, APPLE, AND IBM—AND MARKETS FOR RADICAL INNOVATION[†]

8.1.1 Problems of Technology Leadership—Xerox, Apple, and IBM

Technology leadership is exerted through "function," in producing products with advanced performance or features. Marketing leadership is exerted through "application," in service, distribution, and pricing. Both technology leadership and marketing leadership are important to ensure that the commercial benefits of innovation will be captured by the innovating corporation. An example of the advantages and perils of technology leadership can be seen in the early 1980s in the experience of three competitors in the personal computer and office automation market: Xerox, Apple, and IBM.

We have seen that personal computers emerged as a mass market by 1980. Then, in 1984, the second generation of personal computers was introduced, focusing on office systems and user-friendly technology. "Office systems" consisted of networks in which personal computers served as distributed workstations. "User

---

[†]This case study is adapted from F. Betz, *Managing Technology*, © 1987, pp. 113–116 and 118–119. Reprinted by permission of Prentice-Hall, Inc., Englewood Cliffs, NJ.

friendliness" consisted of the integration of applications software and improved human/computer interfaces (such as higher-resolution screens, iconic menus, and multiwindows). Continuing incremental innovation created the second generation of personal computers. The user-friendliness innovations gained public attention in 1983, when Apple Computer introduced a new personal business computer called Lisa, and in 1984, when it introduced a less-expensive version called the Macintosh. Yet the technology story behind this second generation (e.g., Lisa and Macintosh) began not in Apple but in Xerox.

Up until the early 1970s, Xerox had been only in the reprographic business, grown from the small Haloid company, on the basis of Carlson's xerography patents. However, in 1970, anticipating competitive pressure in the copier market in the long term, Xerox had a new vision. Xerox formulated a long-range plan of office automation as a coming technological revolution. Xerox planned to expand its products to encompass the totality of office systems. Copiers would be a key component, but the whole of the system of office practice would be the product domain. To research and develop products for this new business strategy, Xerox built a new corporate research laboratory, in addition to the older research laboratory in Rochester, New York, which was to continue being devoted to reprographic research (Uttal, 1981).

George Pake, a physicist, was hired to head research and development for Xerox and to build the new laboratory. He chose to locate it in "Silicon Valley" next to Stanford University. Stanford had become the major center of the semiconductor electronics industry in the 1970s. Stanford University's research and educational capability in computers and electronics would provide a stimulating intellectual environment for Xerox's new research laboratories efforts. The new laboratory was named PARC—Palo Alto Research Center: "On a golden hillside in sight of Stanford University nestles Xerox's Palo Alto Research Center, a mecca for talented researchers—and an embarrassment. For the $150 million it has lavished on PARC in 14 years, Xerox has reaped far less than it expected. Yet upstart companies have turned the ideas born there into a crop of promising products. Confides George Pake, Xerox's scholarly research vice president: 'My friends tease me by calling PARC a national resource' (Uttal, 1983, p. 97).

PARC technologically pioneered the new concepts of office systems, of user friendliness, and of distributed computer networks.

In the late 1960s, an MIT professor, J.Licklider, had envisioned an easy-to-use computer. PARC researchers, influenced by Licklider's vision, determined to make it practical. PARC's Alan Kay even projected Licklider's ideas further as a machine that was to be portable, about the size of a book, and easy to use (Bantimo, 1984).

The PARC scientists developed a system called Alto, installing it in PARC. Alto had several unique features. For example, it used a new programming language called Smalltalk, facilitating programming with windows, icons, and "mouse" controllers. Alto was easy to use with an iconic menu pointed to by a movable desktop "mouse" (for example, to file a report or program, the mouse was used to move an arrow on the screen to point to a tiny image of a file drawer). Alto also had multiple window displays (so one could look at several applications at once), called "desktop" organization display. Altos were also hooked up in a network (Ethernet) of communicating workstations. Each researcher had an Alto on his or her desk, sending mail on the computer to one another and printing on a common Xerox laser printer. Alto was the first experiment in the concept of the paperless, automated office.

All this had occurred in the late 1970s at PARC, even before the world realized that a personal computer revolution was brewing in the hobbyist world of computer kits. Xerox's office products division, however, focused on competition in the dedicated word-processing market (in which Wang had leapfrogged both IBM and Xerox). Xerox Office Products Division chose then to follow Wang's distributed workstation concept in implementing PARC's vision of office automation rather than the personal computer form. In 1981, they announced a series of products—a smart typewriter, a personal computer, the Star workstation, and a local-area network, the Ethernet. Yet only the workstation and Ethernet incorporated PARC's far-seeing vision: "By the mid-1970s the center [PARC] was hard at work on the Alto. . . .Alto and its software became . . . popular inside Xerox. Product development, however, was the turf of another Xerox group, which was championing a rival machine called the Star" (Uttal, 1983, p. 98).

The Alto could have been sold as a personal computer, but the Star could not, since it only worked hooked to other Xerox equipment. Xerox's office product systems management appeared not to appreciate the complete set of ideas in Alto and Smalltalk. However, a rival group did: "Lisa is the unkindest cut of all. In December 1979, Steve Jobs . . . visited PARC with some colleagues to poke around. They saw Smalltalk . . . . 'Their eyes bugged out,' recalls Lawrence Tesler, who helped develop Smalltalk. . . . Seven months later Jobs hired Tesler, having decided to use many Smalltalk features in the Lisa" (Uttal, 1983, p. 100).

Xerox's Office Product Division launched a too-expensive and too-closed workstation instead of an advanced personal computer. The personal computer it marketed was a mundane, look-alike personal computer, which did poorly even in the booming market of personal computers. It turned out to be a case of a technology leader without marketing flexibility. The Xerox Star workstation was revolutionary enough to show PARCs vision—to the competition. Apple followed two years later first with its Lisa and then with its Macintosh personal computers. Software for the IBM, Lotus 1-2-3, followed quickly after Lisa. They all had the user-friendly visions of PARC. In sequence, then, MIT inspired Xerox, Xerox inspired Apple, and Apple inspired software companies for IBM.

Apple's president, Steve Jobs, understood the market. It was personal computers that were to lead the way into office systems. He had seen the technical significance of PARC's advances and targeted them toward the right market, the personal computer market. Xerox's Office Products Division had failed to envision the impact of the personal computer on the office system. Technology management requires correct strategy on two sides—technology and market. In reporting on the Xerox-Apple innovations, Bro Uttal (1983) commented: "From this Xerox might appear to have muffed the chance to make it big in personal computers with PARC's creations.... To mourn the Alto, though, is to blame unfairly those who killed it. Xerox was out to produce office equipment, and no office equipment supplier, including IBM, foresaw that personal computers would compete with their wares" (p. 100).

Flexibility in marketing strategy for new-technology products is very important. When the personal computer emerged, Xerox might have made use of it as a workstation, which is eventually how it was used. The PARC scientists had correctly anticipated the decentralized network of computers. This approach might have been used as an alternative at Xerox. Committing too many eggs into one market basket early in a new technology is costly if later it turns out to have been the wrong basket.

8.1.2 Markets for Radical Innovation

Market positioning for radically new innovations usually begins with the niche markets receptive to new technologies. These markets are the military, scientific, hobbyist, and technically oriented industrial markets. The military market is highly organized to fund new technology and new applications. The scientific market is trained to create and improve new instruments. The hobbyist market enjoys learning and views technical tools as a pleasurable avocation. Technically oriented sections of industrial markets (such as engineering design) have technically trained personnel to adopt and improve technical applications. These markets are the first to adopt new high-technology products. They are the traditional markets for small high-tech companies.

To move high-technology products into the mass markets of consumers or of general business, a company must be both a technology leader and a marketing leader. The reason is that mass markets are sensitive to both education and price. They require substantial training and service to accept an innovative product. They also require a relatively low price, high level of safety, and ease of use.

In a mass market, a technology leader runs the risk of demonstrating a new market, which can later be captured by a technology follower with stronger marketing and lower-cost production capability—the "me-too" competitor. The advantage to the "me-too" company is that the potential market factors for success have been explored by the technology leader and thereby delineated. The competitor can then improve its product, more carefully target and price the market, and provide features and service, which together may cancel the lead-time advantage of the technology leader.

8.1.3 IBM PC

In computers, marketing leadership has gone to the company that had hardware performance almost as good as anyone else's but provided superior software for applications. With this principle, IBM dominated the mainframe computer market from the 1960s through the 1980s. In the new personal computer market at the beginning of the 1980s, IBM had watched Apple and the growing personal computer market, then attacked with the IBM PC. Introduced in 1981, it claimed 7% of the market that year. In 1982, it matched Apple's share at 27%. In 1983, IBM emerged as the clear and dominant leader with 36% and Apple had slipped even further, to 24%.

IBM played a "close-follower" technology strategy in 1981 in entering the personal computer market. It used just the right balance of technology leadership and marketing leadership—a little technology leadership and a lot of marketing leadership. The IBM PC had a 16-bit central processing chip, superior to the 8-bit chip of the Apple and competitors in the personal computer market in 1981 (superior in the sense that it allowed a larger memory address). Yet since the IBM PC used an 8-bit bus, the combination made it easy for software applications to be written for the IBM PC (using the larger memory that the larger addressing capacity of the 16-bit processor made possible). Therefore, the principle of IBM's entry into the personal computer market in 1981 was just a little technology leadership, but not too much.

One marketing aspect of IBM's successful entry into the personal computer market in 1981 was that IBM focused on the business market, pricing it right for that range and marketing it through its own sales personnel, opening new retail outlets, and distributing it through the new personal computer dealer networks. The important marketing edge was the IBM name, the installed base of IBM business customers, and IBM's reputation for service. This combination of the right balance of marketing and technology leadership worked for IBM. There was an IBM for small businesses and for sharp business executives who knew that IBM mainframes served their businesses. Another factor for success was the choice of an open architecture and operating system (imitating Apple), which facilitated the rapid transfer and development of software applications for the new IBM personal computer.

It was to survive IBM's attack that Apple determined to seize technology leadership in 1981. Apple rushed through the Lisa project to grab technology leadership from IBM (and Apple had found the ideas for technology leadership at Xerox). In 1983, Apple introduced the Lisa to compete with IBM—but met with disappointment. While recognized as a technology leader, Lisa had been priced too high at $10,000, almost twice the cost of the IBM PC XT (Morrison, 1984).

Moreover, some of the software advantages, the integrated programs, had not been strong enough to knock IBM out immediately—because "'me-too" software products for the IBM PC quickly began to appear. Once such program became the best selling software in the middle 1980s, Lotus 1-2-3. This program had copied one of Lisa's program integration features (database and spreadsheet), yet ran on IBM PCs, which by then dominated the market.

Apple responded in 1984 by introducing the Macintosh, a smaller version of the Lisa and priced in the IBM PC range, $2000 to $3000. This model sold well initially (in contrast to the Lisa model) and kept Apple as a major competitor to IBM in the personal computer market in the middle 1980s. Apple had made the strategic choice of playing technology leader against IBM's market leadership. Without IBM's prestige, mainframe business, and extensive business installations, Apple could not then match IBM's marketing strength. Wars in business competition are fought battle by battle, in products pitted against each other in technology, marketing, and pricing.

8.2 CASE STUDY TWO: COMPUTER-INTEGRATED MANUFACTURING IMPLEMENTATION AT A GENERAL ELECTRIC PLANT†

8.2.1 Introduction

The GE Electrical Distribution & Control business (ED&C) focuses on supplying electrical power distribution and controls the marketplace with high-quality products and superior service. ED&C competes in tough domestic and international markets and has a strategy of not only maintaining, but increasing, its leading market share. ED&C's approach to meeting competitive challenges and winning increased share is through *total automation*—a focus on automating all functions of the business, not just production, for total productivity. One element of the overall strategy is aggressive new product introductions coupled with state-of-the-market Manufacturing Centers of Excellence for low-cost production.

Total automation is more than just equipment. It involves the total technology management from product definition via CAD/CAM technology to order entry, materials procurement, equipment and process development, quality control, process planning, production, shipment, and finally invoicing. It involves the development and integration of multiple computer-based systems into a master control network, which forms a framework and foundation from which a business launches its cost and product leadership objectives. Establishing these integrated systems has permitted improved customer service from the initial proposition phase of an order to short-cycle delivery of product to precise customer specifications via a sophisticated CIM Manufacturing Center.

The GE ED&C Manufacturing Center of Excellence that best reflects the results of a total automation effort is the manufacturing plant in Salisbury, North Carolina, where electrical lighting panelboards are produced.

8.2.2 A Lighting Panelboard

A lighting panelboard is a device that provides four basic functions to the end user: It (1) provides electrical service disconnection, (2) distributes the energy in a building by breaking the service up into circuits, (3) provides overcurrent protection and short circuit protection, and, of course, (4) provides for the safety of people and property.

It looks very much like the load center in homes that brings in electric power and breaks it up into numerous protected circuits. Lighting panels typically carry more power than a load center, but the function is about the same.

Lighting panels are used in high-rise residential applications, or in industrial and

---

†This case study was prepared by William J. Sheeran, Thomas R. Campbell, and Jill C. Sutton, of the General Electric Distribution and Control Center, Plainville, CT. The material is included with permission.

commercial environments, and are built for a wide range of applications, some very simple and some very complex. Indoor lighting, outdoor lighting, timing, switching, and relaying functions all represent application of lighting panelboards.

8.2.3 Summary: A-Series Lighting Panelboard Project

This aggressive program involved the redesign of an existing lighting panelboard product from a customized (28,000 unique parts) and costly product manufactured at six domestic locations in which each order was requisition engineered with an overall order/build/ship cycle of three weeks to a streamlined product (fewer than 1300 unique parts) produced in one location with no requisition engineering needed and a two-day order/build/ship cycle time capability.

The Salisbury, North Carolina factory operates with one-sixth the number of people required to produce the old style panelboard product and contains the latest state-of-the-art technology for sheetmetal fabrication and automatic assembly. The total investment in this operation exceeds $11.0 million, with an annual benefit in excess of $5.0 million being realized—a 30% cost reduction.

At the core of this automated factory is a unique Flexible Fabrication System (FFS) for sheetmetal fabrication (see Figure 8.1). The system contains 13 machine tools, which are linked by a variety of automated material-handling units and controlled by a central supervisory computer, which directs the processing of over 7,000 tons of raw sheet metal through four integrated fabrication lines. These fabrication lines incorporate punching, bending, embossing, shearing, and forming operations under complete programmable computer control and represent the largest and most complex systems of its kind in the world today.

Production orders (see top of Figure 8.2) for parts to be produced are communicated nightly to the FFS computer from an automated order entry system located in Plainville, Connecticut, which collects *all* field panelboard orders from around the country and automatically requisitions engineers and verifies them for accuracy. The FFS computer then optimizes the orders to maximize material utilization and minimize processing time for production of lighting panel enclosures.

Additional fabrication and assembly cells are also in place to support the construction of the interiors (busbars, rails, circuit breakers, miscellaneous hardware) of the lighting panels. A Busbar/Z-Rail cell produces parts on a just-in-time basis as demanded by an automated Interior Assembly cell. The Busbar/Z-Rail cell contains three fabrication lines regulated by GE Series-Six Programmable Controls and lined through a GE Workmaster console. The cell produces the components required by the Interior Assembly cell and maintains a small (work-in-process) queue ahead of the cell.

The Branch Interior Assembly Cell then assembles lighting panel interiors via two robotic-based workstations linked by an indexing conveyor. The cell assembles a variety of small detail parts and GE plastic components, automatically using four GE A-12 robots regulated by RC 1000 controls linked to a bank of GE Series-Six programmable control and GE Workmaster workstations. The cell assembles interiors at a rate of 240–260/hour.

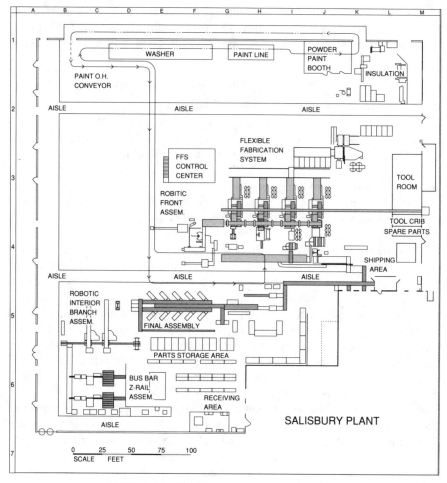

FIGURE 8.1 Flexible Fabrication System (FFS).

8.2.4 Product Design for CIM

As has been mentioned, the goal of this project was to design and produce a product that has a one- to three-day customer response cycle. The existing cycle time was two to three weeks, a big challenge.

On day 1 the order is placed, via an automated requisition engineering system that maintains standardization of design as demanded by our structured product line, and sent via computer network to Plainville, Connecticut. The IBM System 38 sorts and prioritizes the orders, defines the materials, orders the parts, and determines the configuration of the end product. At the end of the first day, the orders are downloaded to the FFS System's PDP11/44 in Salisbury to prepare for production. On day 2 the orders that were downloaded overnight are produced. The system produces in varying lot sizes, with minimum inventory, minimum

CASE STUDY TWO: COMPUTER-INTEGRATED MANUFACTURING IMPLEMENTATION

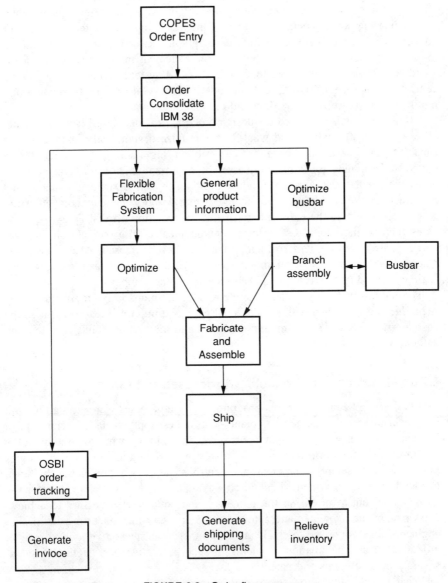

FIGURE 8.2 Order flow process.

work force, and minimum cost. The orders are then packed and loaded on a truck. On day 3 the order is on its way to the customer.

In order to define what the product had to be, we conducted detailed marketing, manufacturing, requisition engineering, and automated proposition system studies. We looked at competing products and manufacturing methods to gain what insight we could and developed strategies for improving product design for manufacturing while further increasing the marketplace features.

Many computerized designing tools were applied to optimize the design while minimizing the development costs. One of the key computerized tools is used for interference detection between mating parts, graphic assembly and test simulation to minimize the number of prototypes, and finite element analysis to evaluate stress and deformation behavior under load. The CALMA system also created numerical control output for machining, and supported the development of fabrication and mold tooling, as well as moldflow and thermal stress analysis.

All of these tools shortened the design process, provided a basis for conducting efficient trade-off analysis and what-ifs to optimize design, and provided design capability that would not have been possible using traditional product design methods. Designing for simplicity, assembly, and automation go hand in hand. What we needed was a single product design that was designed for automated manufacture. We couldn't just automate existing designs, as many companies did so unsuccessfully in the early frenzy of U.S. automation efforts. Parts had to be redesigned so they could be assembled automatically. The overall solution was less expensive, but the part cost and count may have increased. Also, constraints placed on us by Underwriter Laboratories (UL) and the National Electrical Code at times caused us to change and even redesign some components.

The end result of this constant struggle was the optimum design for manufacture. A product was designed with standardized parts and improved features that fit the customers' needs while minimizing total cost and improving productivity through automated production.

8.2.5 Manufacturing Technology/Processes and CIM

In 1984, at the beginning of this program, many businesses were struggling with how to change their manufacturing plants so as to best apply the latest technological innovations in integrated computer-assisted manufacturing while migrating to the just-in-time methods of production. We also struggled with these new technologies in fabrication, assembly equipment, and control systems. We took on the challenge of identifying the best methods required—available or not, proven or unproven—and then set out to provision the processes and equipment necessary to achieve our vision of the integrated and automated factory. Our manufacturing and design engineers alike continued to challenge each other in an interactive loop to resolve producibility and automation problems in such a creative manner that it led to several novel solutions and patent awards.

What, specifically, drove us to new manufacturing technology? Certainly, a new streamlined design represented a challenge in that it contained fewer parts, but many new parts were now in GE engineering plastics which was relatively new to us. Could we achieve the tolerances demanded by the design? How would these parts react to handling and automated assembly equipment? These new demands challenged our technical skills and forced us to again be creative and resourceful in seeking clever cost-effective solutions.

Product designing was only one element of the key drivers that led to our contemporary factory. No inventory, minimum WIP, and lot size of one forced us to totally change the way we manufactured and the kind of equipment and methods we adopted. A backdrop of the ever-increasing notoriety of JIT—punctuated by the cry of "Automate, Emigrate, or Evaporate"—drove us even harder to achieving the maximum productivity possible from our manufacturing technologies. The manufacturing processes employed are not new—sheetmetal punching, shearing, bending, embossing, stamping, and roll forming. What *is* new is the way we have applied and integrated some of the latest in computerized equipment technology and tooling methods to our product. We set out to find strong, capable vendors who were willing to work closely with us as we pushed the state of the art in sheetmetal fabrication and assembly automation technology. In 1984 there were no fully computer-controlled and -integrated sheetmetal flexible manufacturing systems (FMS) to be purchased. It had to be developed, so we were going to develop it. We searched especially hard to find a sheetmetal fabrication vendor who was willing to take on the challenge—and who had the machine tools and software development resources necessary and the know-how to alter our designs where required to achieve a producible product. We found that vendor is Sarago, Italy, and their FFS forms the core of the Salisbury operation.

In assembly, however, we certainly have done a lot to advance the state of the art in design for assembly methods and robotic application. We were very ambitious in what we set out to automate; we worked closely with design engineers during the product design phase to guarantee that the top-down assembly approach was being maintained. We did, however, exercise some caution and automated only up to the point of extreme diversity; our final assembly area, for example, is manual.

8.2.6 FFS Control System

The FFS Control System architecture is based on a Digital Equipment Company PDP 11/44 supervisory computer, which is linked via leased line to an IBM System 38 in Plainville (see Figure 8.3) to accept order data automatically on a daily basis. The control system software performs its optimization tasks and produces a two-shift schedule for each day for shearing material and distributing to the fabrication lines in a balanced and uniform loading sequence. Real-time communications are constantly maintained with PDP 11/73 line controllers (see Figure 8.4), which control all machine and line functions.

Each line controller is a powerful computer resource in itself, and handles all local machine-specific tasks as well as peripheral equipment serving the lines. Error diagnostics and other messages are reported to line operators as needed with a constant dialogue and part-tracking subsystem remaining active at all times. High-speed data communications are maintained using standard digital RSX11M operating features with customized RS-232 serial links via HS modems. The system also has the capability to create management reports as requested, based on any snapshot in time.

8.2.7 Plant Operations and Work Force

The Salisbury, North Carolina plant represents not only the focal point for GE ED&C's application of the latest in CIM technology, it also operates using a

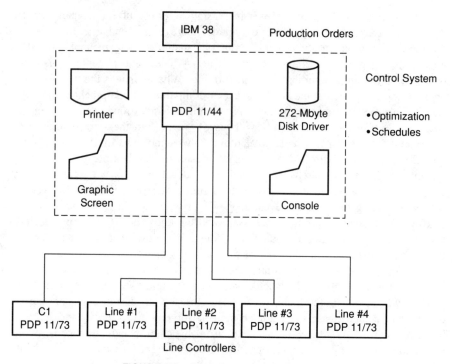

FIGURE 8.3 Control systems configuration.

novel "self-directed work force concept"; that is, a workforce organization where the traditional line supervisor and foreman structure is eliminated and ownership for the plant operation, product quality, and customer service is distributed to *all* plant employees. The result is that everyone gets involved in developing, refining, and improving both product and process a continuing basis with a constant focus on process efficiency, product quality, and customer service—not just high-tech automation of a product.

8.2.8 Payoff

And what were the results of this total automation effort?

- Lighting panelboard plants reduced from six to one.
- Customer delivery cycle reduced from two weeks to three days.
- Total cost reduction of 30%.
- Direct labor reduction of 55%.
- Elimination of requisition engineering.
- Inventory reduced from 45 days to 9 days.

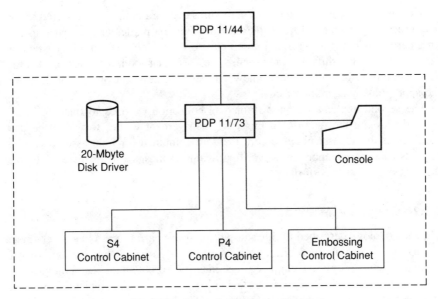

FIGURE 8.4 Line controller configuration.

This project has been an outstanding achievement for GE-ED&C—one that has allowed us to fully realize our vision of CIM and the modern factory while keeping our business at the state of the market.

In closing: What is possible with CIM and what is needed to realize it both start with product design for CIM, not the application of CIM to existing designs. CIM can have positive impact if you realize the real sequence necessary to achieve it.

8.3 CASE STUDY THREE: CHOOSING AMONG COMPETING TECHNOLOGIES—A STUDY OF AUTOMATIC IDENTIFICATION TECHNOLOGIES[†]

8.3.1 Introduction

The rate of growth for new technology is expanding as it never has before. This rapid change can often leave managers and engineers confused as to which technology best suits their purposes. To tackle the problem of choosing among a number of new technologies, a methodology that provides a means to make a rational decision is proposed. The case study then uses the proposed methodology by choosing an automatic identification technology for two common problems in plant-floor data processing of time and attendance and shop-floor data collection. These two examples are simplified applications of real problems faced in plant-floor data collection. The purpose of their inclusion is to demonstrate how the methodology is used to help choose among competing technologies.

[†]This case study was prepared by Scott McConnell. Entry Systems Division of IBM Corporation, Boca Raton, FL. The material is included with permission.

The time and attendance application requires that information be collected so the data-processing center can track the amount of time a particular employee spends on a particular job. For instance, an employee will "clock on" to job A when he begins work on his shift. He may then "clock off" that job after a period of time. The data-processing center must know the actual time spent on the job to make the appropriate accounting of charges.

The second application, shop-floor data collection, is used to track the flow of work through the factory. In this case management wishes to verify the part numbers of the subassemblies before they are combined for final assembly. This requires each part number to be verified against a master bill of material before the next operation is performed.

8.3.2 Evaluation Methodology

The methodology described in this study has a four-phase process that an engineer can use to choose an appropriate technology for a particular application. Before using the methodology, the engineer must identify a problem and technologies that could be used to solve it. The list of technologies must be as complete as possible and, therefore, should include any technology that has a remote possibility of being used.

Phase 1: Attribute List. Every technology has positive and negative attributes. This phase of the process is the identification of these attributes by the engineers doing the research. They should gather as much information as possible on each technology under consideration from sources such as technical bulletins, sales literature, and people knowledgeable in the field. From the investigation the engi-

Application Evaluation		Device 1	Device 2	Device 3	Device 4	Device 5
2	Attribute 1	5 / 10	3 / 6	0 / 0	4 / 8	2 / 4
0	Attribute 2	2 / 0	5 / 0	3 / 0	4 / 0	2 / 0
1	Attribute 3	4 / 4	1 / 1	2 / 2	1 / 1	3 / 3
5	Attribute 4	1 / 5	0 / 0	1 / 5	5 / 25	1 / 5
Total		19		7	34	12

FIGURE 8.5 Sample decision matrix.

neers can generate a list of characteristics taken from all the technologies under review. This attribute list is the base of the methodology and must be as complete as possible. The attributes can then be listed in a column of the decision matrix, as shown in Figure 8.5.

Phase 2: Technology Ratings. The next step is to evaluate each methodology based on its relative strength to the others for each of the attributes. The rating system used is:

5 excellent technology for this attribute
4 above average
3 average
2 below average
1 poor
0 technology does not possess this attribute

The ratings for the technologies are placed in the upper portion of the cells in a matrix underneath the technology. For the example shown in Figure 8.5, technology 2 is the strongest technology for attribute 2, but does not possess attribute 4.

Phase 3: Application Evaluation. The third step in the procedure switches the focus from the technologies to the problem being solved. The engineers generate a short description of each of the attributes listed in Phase 1 describing what the attribute is, why it is important, and what it could mean to the user. Using these descriptions, a separate team of engineers and managers evaluates the need of each attribute in solving the problem. Again, a relative rating system is used:

5 must have this attribute to solve problem
4 extremely important, but could adjust without it
3 important
2 not important, but would be beneficial
1 do not need
0 do not want to consider in the decision

The values assigned by the team for the attributes are listed along the left side of the matrix under the column called Application Evaluation. In Figure 8.5, attribute 4 is required for the problem, and attribute 2 will not be included in the decision.

Phase 4: Decision Matrix Calculation. After the application requirements are rated, the engineers may now complete the decision matrix and make recommendations based upon its results. First, any technology with a 0 value for an attribute (upper portion of cell in matrix) that is required by the application (5 in the Application Evaluation column) is eliminated from the decision matrix. This is shown in Figure 8.5 for Device 2 (attribute 4). The remaining technology columns are computed by multiplying the upper portion of the cell under each technology column by the number assigned at the left under the Application Evaluation column. The result is recorded in the lower portion of the cell. These results are then summed by column and recorded at the bottom of each Device column. The technologies with the highest ratings should be most seriously considered for implementation. Figure 8.5 shows the calculations, which result in choosing technology 4 as most appropriate for the sample application.

8.3.3 Selection among Automatic Identification Technology Alternatives

As the competition for market share in many of the maturing industries increases, the battleground for the market will increasingly be fought with price. This increased emphasis on lower price is also increasing the emphasis on productivity enhancements in our workplaces. The low-cost producers will ultimately control the mature industries and force the less productive firms out of the game. One area of work that has been receiving increased attention from productivity-minded engineers and managers is data collection. Data collection is used for functions such as inventory control, production scheduling, time and attendance, shop-floor control, and other data-intensive functions. The productivity enhancements are derived from two main areas. First, the rate at which an average worker enters 12 characters through a keyboard data entry device is 6 seconds compared to entry rates ranging from 4 to 0.1 seconds for automatic entry devices. The other area that enhances productivity is the reduction of costly input errors. The average worker will make 1 substitution error for every 300 characters entered from a keyboard. Error rates range from one in 10,000 to virtually none for automatic data entry devices, so substantial savings may be realized in certain applications from this fact alone (Adams, 1986).

The following is a list of the automatic identification technologies chosen for the evaluation:

- Voice recognition
- Magnetic slot reader
- Bar code wand reader
- Hand-held optical character recognition reader
- Optical character recognition page reader
- Active radio-frequency tag
- Passive radio-frequency tag
- Holographic bar code scanner
- Fixed-beam bar code scanner
- Hand-held laser bar code scanner

Phase 1: Attribute List

- Price of reader
- Substitution error rate
- Price of media
- Human readability of media
- Reading distance flexibility
- Line of sight requirement
- Life of media
- Reader's resistance to harsh environment
- Media's resistance to harsh environment
- Ability to alter media
- Density of data
- Security of media
- Operator requirement

Application evaulation		Voice	Magnetic slot	Wand	OCR hand held	OCR page	RF active	RF passive	Holographic	Fixed	Laser
2	Price of reader	2/4	5/10	5/10	4/8	3/6	4/8	4/8	1/2	2/4	3/6
4	Substitution error rate	1/4	4/16	4/16	2/8	2/8	5/20	5/20	4/20	4/16	4/16
5	Price of media	5/25	3/15	4/20	4/20	4/20	1/5	1/5	4/20	4/20	4/20
0	Human readability	5/0	0/0	3/0	5/0	5/0	0/0	0/0	3/0	3/0	3/0
0	Reading distance flexibility	2/0	1/0	1/0	1/0	1/0	5/0	4/0	3/0	3/0	3/0
0	Line of sight requirement	4/0	0/0	0/0	0/0	0/0	5/0	5/0	4/0	0/0	2/0
4	Life of media	5/20	2/8	3/12	3/12	3/12	4/16	5/20	3/12	3/12	3/12
4	Reader's resistance to harsh	2/8	3/12	4/16	3/12	3/12	4/16	4/16	4/16	4/16	3/12
4	Media's resistance to harsh	2/8	3/12	1/4	1/4	1/4	5/20	5/20	1/4	1/4	1/4
4	Ability to alter media	5/20	5/20	0/0	0/0	0/0	5/20	0/0	0/0	0/0	0/0
4	Density of data	3/12	5/20	3/12	1/4	1/4	5/20	4/16	3/12	3/12	3/12
5	Security of media	1/5	5/25	4/20	1/5	1/5	5/25	5/25	1/5	1/5	1/5
0	Operator requirements	3/0	0/0	0/0	0/0	3/0	5/0	5/0	5/0	0/0	0/0
3	Speed in reading data	1/3	2/6	2/6	1/3	1/3	5/15	5/15	4/12	4/12	4/12
3	Proven technology	1/3	5/15	5/15	4/12	4/12	1/3	1/3	4/12	5/15	5/15
	Total	112	159	131	88	86	168	148	111	116	114

FIGURE 8.6 Time and attendance decision matrix.

User		Voice	Magnetic slot	Wand	OCR hand-held	OCR page	RF active	RF passive	Holographic	Fixed	Laser
5	Price of reader	2/10	5/25	5/25	4/20	3/15	4/20	4/20	1/5	2/10	3/15
4	Substitution error rate	1/4	4/16	4/16	2/8	2/8	5/20	5/20	4/20	4/16	4/16
5	Price of media	5/25	3/15	4/20	4/20	4/20	1/5	1/5	4/20	4/20	4/20
3	Human readability	5/15	0/0	3/9	5/15	5/15	0/0	0/0	3/9	3/9	3/9
0	Reading distance flexibility	2/0	1/0	1/0	1/0	1/0	5/0	4/0	3/0	3/0	3/0
0	Line of sight requirement	4/0	0/0	0/0	0/0	0/0	5/0	5/0	4/0	0/0	2/0
2	Life of media	5/10	2/4	3/6	3/6	3/6	4/8	5/10	3/6	3/6	3/6
4	Reader's resistance to harsh	2/8	3/12	4/16	3/12	3/12	4/16	4/16	4/16	4/16	3/12
3	Media's resistance to harsh	2/6	3/9	1/3	1/3	1/3	5/15	5/15	1/3	1/3	1/3
0	Ability to alter media	5/0	5/0	0/0	0/0	0/0	5/0	0/0	0/0	0/0	0/0
3	Density of data	3/9	5/15	3/9	1/3	1/3	5/15	4/12	3/9	3/9	3/9
0	Security of media	1/0	5/0	4/0	1/0	1/0	5/0	5/0	1/0	1/0	1/0
0	Operator requirements	3/0	0/0	0/0	0/0	3/0	5/0	5/0	5/0	5/0	0/0
4	Speed in reading data	1/4	2/8	2/8	1/4	1/4	5/20	5/20	4/16	4/16	4/16
4	Proven technology	1/4	5/20	5/20	4/16	4/16	1/4	1/4	4/16	5/20	5/20
	Total	95	124	132	107	102	123	122	116	125	126

FIGURE 8.7 Shop-floor collection decision matrix.

- Speed in reading data
- Proven technology

Phase 2: Technology Ratings. The matrices in Figures 8.6 and 8.7 show the values assigned to the technologies during the technology review.

Phase 3: Application Evaluation. As described earlier, two theoretical applications were chosen to implement the methodology. The first is a time and attendance application used to monitor actual work time spent by employees on a particular job. The data is used to make appropriate charges to the jobs. The decision matrix for this application is shown in Figure 8.6. The second application was a shop-floor data entry application in which the management of a plant wished to control the flow of work through the factory. This matrix is shown in Figure 8.7.

The following are descriptions for the attribute list generated during phase 1.

1. Price of reader: How important is the price of the reader? To answer this, consider the number of installations that you will need for the application.

CASE STUDY THREE: CHOOSING AMONG COMPETING TECHNOLOGIES 161

For instance, if your application needs 10,000 readers, the cost may be more important than if it needed five readers.

2. Substitution error rate: How critical is data integrity? For an application that will be verified several times before the data is processed, substitution errors may not be as critical as in those that are processed immediately. A substitution error is an error that is not detected by the reading device.
3. Price of media: The medium is the object being read by the identification device. For bar code the medium is a paper label. The cost of media increases in importance as the number of objects to be identified increases. If a tag is being placed on every component that enters a manufacturing plant, then the cost of those tags is more important than if they were being used to track tools.
4. Human readability of media: Some identification media cannot be read by a human. If the application needs to have a medium that can be read by an operator, then high importance should be assigned here.
5. Reading distance flexibility: Is there a problem in your application that does not permit the object being read to come in contact with the reader? A fork lift driver may be more productive using a bar code laser than a bar code wand since the laser can read labels from a distance. This would enable the drive to stay on his truck and read labels from the pallet. In an application where a document is being passed by a workstation, a contact reader may not cause any inconvenience.
6. Line of sight requirement: Does the application require a reader to read the identification tag around a corner or inside of a box? Some applications cannot ensure that the identification tag will be facing the reader. If this is important in your application, assign a high level to this attribute.
7. Life of media: Will the identification tag be used repeatedly or will it be used once and discarded? Some devices do not lend themselves to multiple readings.
8. Reader resistance to harsh environment: How harsh is the environment in which the technology will be required to function? Is there a temperature or noise constraint inherent in the environment? If so, assign a high level of importance to this attribute.
9. Media's resistance to harsh environments: Will the object being identified travel through processes that will subject it to harsh environment such as temperature or humidity? Some technologies' media are more resistant than others to the environment.
10. Ability to alter media: Do the media need to be altered during the manufacturing process? For instance, if an object is being used to identify quantity, does it need to be rewritten after a piece is used?
11. Density of data: How much information needs to be encoded on the object? Can a coding system that provides five characters suffice or do you need three paragraphs of description?

12. Security of media: Does the application require that the media be read only by authorized personnel? Is it critical that an identification tag not be copied by a copy machine?
13. Operator requirement: Does the application require the operator to enter data and perform other tasks simultaneously? If so, assign a high level of importance here.
14. Speed in reading data: How important is speed to your application? Will the operation require reading four tags an hour or four tags a second? The more important speed is, the higher the rating for this attribute.
15. Proven technology: This attribute is a measure of aversion to risk. Some of the technologies in this evaluation have been around for years, whereas others are just now gaining popularity. If lack of risk is important, assign a high rating to this attribute. If you do not want this considered in the evaluation, assign a zero.

Phase 4: Decision Matrix Calculation. The decision matrix results are shown in the following Figures 8.6 and 8.7. For the time and attendance application, the chosen technology is either active radio-frequency tags or magnetic slot readers. The shop-floor control application suggests the bar code wand is the primary technology.

8.3.4 Summary and Discussion

Evaluation of new technologies is a complex process facing engineers and managers of both small and large companies. The development and use of objective, systematic methodologies to aid them in their evaluations would greatly enhance the effectiveness of the decision-making process. The result is a more orderly transfer of technology with excellent potential for reducing cost, increasing productivity, and effecting improvement in the company's competitive position. This case study has provided one methodology that can be useful in this regard and has illustrated how this methodology can be used in selecting appropriate automatic identification technology for a specific application.

8.4 CASE STUDY FOUR: ERGONOMIC ISSUES IN TECHNOLOGY APPLICATION[†]

8.4.1 Research Site and Methodology

Data was collected at a manufacturing plant that introduced and used a specific type of computer (J-2000) to assemble printed circuit boards.

[†]Adapted from J. Edosomwan, *1986 International Industrial Engineering Conference Proceedings.* © Institute of Industrial Engineers, 25 Technology Park/Atlanta, Norcross, GA 30092.

8.4.2 Description of the Computer-Aided Task

The computer-aided printed circuit board assembling task was concerned with the use of a logistical application program to assist production workers in the insertion of small electronic components such as transistors, diodes, resistors, and capacitors into a blank printed circuit board.

The J-2000 Component Locator System, which was used in connection with the computer's software, comprised an X/Y table, projector, and the microprocessor driven control. A control panel provided the controls and displays required for the operating system. Included on the control panel were a mini-floppy disk and a keyboard by which random sequence numbers (instructions) were selected. Positioning information for the X/Y table and projector was programmed as machine instructions on the mini-floppy disk.

Additional instructions provided system control functions such as bin advance. The controller distinguished the various kinds of instructions from each other by a descriptor contained within the instruction format. For each kind of instruction, the descriptor was assigned different values.

The process of creating a good printed circuit board assembly program required a considerable amount of forethought. The programmer started by having a completed sample board, an empty board, a drawing that showed component layout, a part list, and samples of each component inserted on the raw card. These items were used to determine the most efficient assembly sequence. Once programmed properly, the logic of the computer caused a light beam to pinpoint a location on the printed circuit board. As the operator pressed the foot switch, the sequence number was incremented by one count, and the instruction residing at that address was fetched and executed. If the instruction was for a component insertion, the X/Y table was moved so that the projector's fixed light spot fell on a location where one end of a component was to be inserted. The projector's movable light spot was also moved so that it fell on the location where the other end of the component was to be inserted. Polarity of diodes and capacitors was also defined by two spots.

Usually, the program began with some rotary bin control instructions (a reset and an advance) so that the components in the various bins were presented to the operator. When a dual in-line integrated circuit was to be inserted, the X/Y moves were preceded by a dip dispenser tube select instruction. This instruction turned a light on to indicate the tube from which a selection was to be made. When the component had been inserted, the operator maintained a gentle downward pressure on the component and pressed the foot switch. Before the next instruction was fetched, the Cut/Clinch mechanism was activated.

The computer was programmed to allow the human operator a certain time interval before moving to the next instruction for component insertion and repeating the same sequence of steps. The programming of the computer consisted of determining the order of components to be inserted and coding that sequence into the memory of the computer. The components also had to be loaded onto reels into the machine. When the machine ran out of parts, an indicator light appeared and the machine remained at interrupt state until parts were replenished. The

TABLE 8.1 Research Sequence

Group	Task Sequence	
	Order 1	Order 2
Group One	Low computer speed Low task complexity Low level of workplace design	High computer speed High task complexity High level of workplace design
Group Two	High computer speed High task complexity High level of workplace design	Low computer speed Low task complexity Low level of workplace design

work flow was also programmed to match the raw card selected for insertion of the components.

8.4.3 The Subjects and Research Sequence

Two groups of five subjects each were studied. One group initially performed the computer-aided task at a specified lower level of task complexity, computer speed, and workplace design. The other group initially performed the computer-aided task at a varied higher level of task complexity, computer speed, and modified workplace design. The order in which the two groups performed the assembly

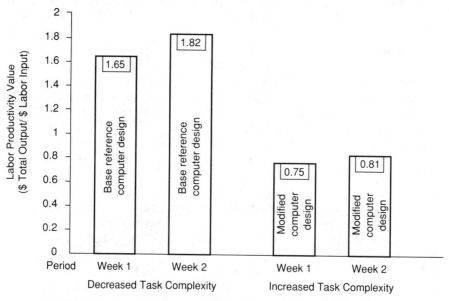

FIGURE 8.8 The impact of computer task complexity on labor productivity.

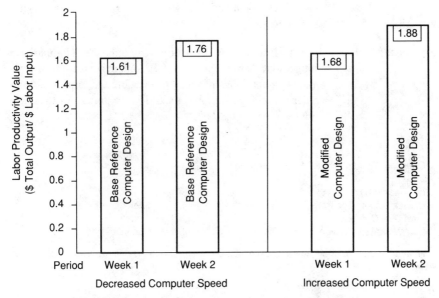

FIGURE 8.9 The impact of computer speed on labor productivity.

task was reversed after a one-week production period. Data on labor productivity and job satisfaction was collected daily for two weeks. The research sequence is presented in Table 8.1.

8.4.4 Measurement Methods for Variables

Productivity: Partial productivity, the ratio of total measurable output to one class of input, was used in this study.

$$\begin{array}{l}\text{Partial Productivity of} \\ \text{task } i \text{ with respect to} \\ \textit{labor} \text{ input at site } j, \\ \text{in period } t\end{array} = \frac{\text{Total measurable output of task } i \text{ performed at site } j, \text{ in period } t}{\text{Measurable labor input of task } i \text{ performed at site } j, \text{ in period } t}$$

Computer Speed: This is the time interval the computer allowed for operators to insert a component into the printed circuit board.

Task Complexity: This is the number of elements (operations) involved in the assembly sequence for printed circuit boards.

Workplace Design: This is measured by an operator-rated acceptable index level for work envelope fixture and system parameters. The index was developed through scale interval (1 = excellent, 5 = poor).

Job Satisfaction: Job satisfaction is measured through the Comparative Judgement Instrument (CJI). CJI contains questions in automated technologies grouped to measure the impact of computer technology on job satisfaction. Examples of the questions are presented below.

Question: Which Task is More Interesting?

( ) Task (C_1^*)
Low Computer Speed
Low Task Complexity
Low Workplace Design

( ) Task (C_2^{**})
High Computer Speed
High Task Complexity
High Workplace Design

Rate each task, to show how interesting it is by writing C_2^* and C_2^{**} in the appropriate spaces below:

| 1 | 2 | 3 | 4 | 5 |
| Not Interesting At All | Somewhat Interesting | Moderately Interesting | Fairly Interesting | Very Interesting |

Question: Which Task is More Stressful?

( ) Task (C_1^*) ( ) Task (C_2^{**})

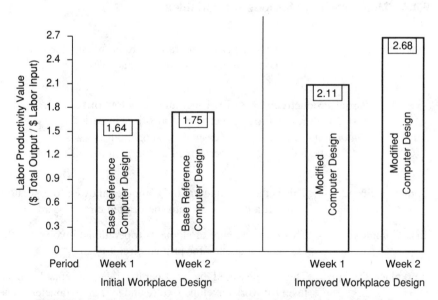

FIGURE 8.10 The impact of computer workplace design on labor productivity.

CASE STUDY FOUR: ERGONOMIC ISSUES IN TECHNOLOGY APPLICATION

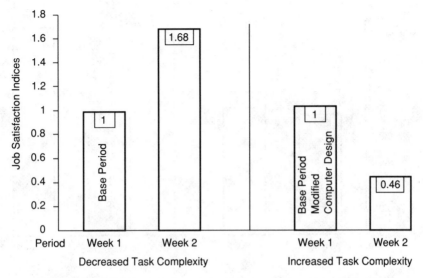

FIGURE 8.11 The impact of computer task complexity on job satisfaction.

Rate each task, to show how stressful it is by writing C_1^* and C_2^{**} in the appropriate spaces below:

1	2	3	4	5
Not Stressful At All	Somewhat Stressful	Moderately Stressful	Fairly Stressful	Very Stressful

$C_1^* =$ Low levels of computer speed, task complexity, and workplace design.
$C_2^{**} =$ High levels of computer speed, task complexity, and workplace design.

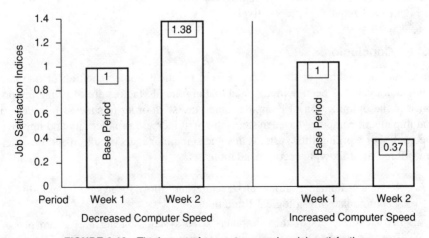

FIGURE 8.12 The impact of computer speed on job satisfaction.

168   CASE STUDIES IN TECHNOLOGY MANAGEMENT

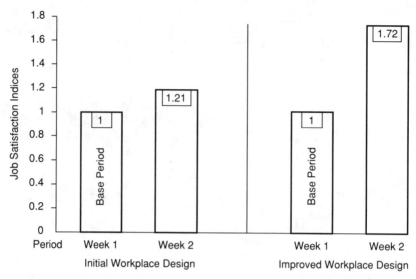

FIGURE 8.13   The impact of computer workplace design on job satisfaction.

8.4.5   Results

At $\alpha = 0.05$ level of significance, there was no order effect between the two groups. Data for both groups was combined. A significant test was therefore performed at the two levels of experimental factors. Results are shown graphically in Figures 8.8 through 8.13. As shown in Figures 8.8 and 8.11, respectively, increased task complexity decreased labor productivity and also had negative impact in terms of production workers' lower job satisfaction. As presented in Figures 8.9 and 8.12, respectively, increased computer speed increased productivity, but had negative impact on workers' job satisfaction. Improved workplace design increased labor productivity and improved job satisfaction; these results are shown in Figures 8.10 and 8.13, respectively.

8.4.6   Conclusions

Based on the findings of this research from practical implementation of a sound methodology, it has been demonstrated that ergonomic factors are of prime importance in the optimization of computer-aided tasks. In order to successfully design and implement computer-based systems that will increase productivity and improve the quality of working life, systems designers and managers of automated factories are offered the following recommendations.

1. Recognize human anatomical structure and allow for systems flexibility to accommodate physiological limitations.
2. Involve the users of the computer-based systems during the design phase. This will provide the opportunity to correct design problems that might affect

productivity and job satisfaction. Factors such as work envelope, system speed, task complexity, and system user friendliness can be addressed before introduction of the system into the work environment.
3. Recognize the workers' right of control over production of work activity.

Failure to consider ergonomic factors in computer-aided tasks can unintentionally take away control from workers. The task performed becomes so routine and repetitious that it lowers morale and job satisfaction. Factors such as computer speed and task complexity should be made variable to accommodate and provide control for the system users.

8.5 CASE STUDY FIVE: TECHNOLOGY IMPACT ON THE QUALITY OF WORKING LIFE

8.5.1 Situational Analysis of Technology Impact on the Quality of Working Life

In order to assess the impact of a specific type of computer (K-3232) and a specific type of robotic device (M-5050) on the quality of working life, some aspects of the technology outcome dimensions were measured before and after introduction and use of a computer and a robotic device in a printed circuit board manufacturing environment. The comparative judgment instrument (CJI) shown in Appendix B was used. Ten operators responded to the CJI instrument during a study period of 12 weeks.

8.5.2 Results and Discussion

As shown in Table 8.2, the response by workers to the two types of technology was significantly different. The assessment provided the basis for examining the

TABLE 8.2 Case Study Results for Manual, Robotics, and Computer Tasks

Variable	Before New Technology Index Manual Task[a]	After Introduction of New Technology	
		Index Robotic Task[a]	Index Computer Task[a]
Job Satisfaction	3.75	3.10	1.67
Task Fatigue	2.1	1.40	4.16
Task Stress	3.50	2.78	3.71
Task Decision Latitude	4.75	1.02	1.82
Psychological Stress	2.80	2.13	3.91
Task Likeness	1.25	4.01	2.98

[a]Significant at $\alpha = 0.05$

aspects of the technology that have negative impact on employees' job satisfaction and other variables. For the specific types of robotic device and computer studied, system designers improved on systems speed, dialogue quality, and other design parameters before approving both technologies for continuous production after the qualification and study period. For example, a variable computer speed was installed instead of the fixed computer speed. This minimized the task stress and fatigue issues.

8.6 CASE STUDY SIX: JUSTIFYING THE USE OF ROBOTICS

8.6.1 Research Site and Methodology

Data was collected at a manufacturing plant that introduced and used J-1515† robotic devices in the assembling of printed circuit boards. Manual insertion of the electronic components was termed difficult and time-consuming. A comparison of the total productivity of the manual and robotic device methods was needed to select the best production method.

8.6.2 Description of the Manual Method of Assembling Printed Circuit Boards

The manual method of assembling printed circuit boards was a process in which the production worker manually inserted components such as resistors, diodes, modules, transistors, and capacitors into an empty circuit board (PCB) or raw card.

The work area was in an approved electrostatic discharge area that consisted of a grounded work bench, chair, tools, and a light box. The manual assembly process started with the verification of paperwork and parts. The component placement list (CPL), picklist, routing, part numbers, engineering change level, serial numbers, templates, and any special instructions were checked before assembly. The foil, which was a blueprint of component insertions, was placed on the lightbox. The foil contained the outline for two boards. The components were manually inserted one at a time, according to the foil. Upon completion, both cards were verified with the CPL and template. If any rework or scrap was to be done, it was performed in a manner similar to that of insertion. When the printed circuit boards were completely assembled and verified, they were manually placed into trays.

8.6.3 Description of the Robotic Device Method of Assembling Printed Circuit Boards

The robotic device method of printed circuit board assembly was a process in which the production worker assembled printed circuits using a J-1515 robotic device and controller. The robot was used to insert rectangular chip packages, varying from

---

†J-1515 is a pseudonym.

0.5 to 1.5 inches, onto a partially populated printed circuit board.

The work area was in an approved electrostatic discharge area that consisted of J-1515 robotic devices and controller, module insertion heads for 0.5 and 1 inch modules, machine base, clinch unit, application control, cabinet, industrial program computer, module feeder, manual card feeder, card carriers, and a set of inspection templates.

The robotics method of assembly began in a manner similar to the manual method, with the verification of the job paperwork and parts. The paperwork is the same as that used in the manual process with the exception of any special instructions unique to the operation of the robotic device.

Once the job paperwork and parts are verified, the operator is ready to begin the setup of the tool. The setup consists of the following steps: (1) ensure that all switches are in the on position as outlined in the operating procedure, (2) obtain the proper workboard holder, and (3) load onto the fixture. Once the fixtures are loaded, the operator is ready to load the PCB assembly program into memory in preparation for assembly.

The program contains the insertion pattern and indicates which part number is to be loaded into the proper input channel. The program then positions the X–Y table into the proper position and guides the robotic arm to the proper channel to pick up a part and insert it into the PCB.

8.6.4 Measurement Method for Productivity

The task-oriented total productivity measurement (TOTPM) model was used in this study. The TOTPMM has wide application for measuring productivity of tasks in both manufacturing and service organizations; it is very similar to the TOTPMM concept introduced in Section 4.7.

Total productivity, defined for task i, at site j, in period t, is equal to the total measurable output of task i, performed at site j, in period t divided by the total measurable input of task i, performed at site j, in period t. The total measurable output is the combined value of finished units, partial units, and other units associated with units produced. The total measurable input is the combined value of labor, materials, energy, robotics, computer, data processing, and other administrative expenses. Total factor productivity is the ratio of total measurable output to the total measurable labor and capital input. Partial productivity is the ratio of total measurable output to one class of input: labor, for example. The input and output components of TOTPM are shown in Figures 8.14 and 8.15, respectively. The values of the output and input are expressed in constant monetary terms of a given base period.

8.6.5 Implementation Methodology

The objective was to compare the partial and total productivities of the manual and robotic device methods of assembling printed circuit boards. The following implementation steps were followed.

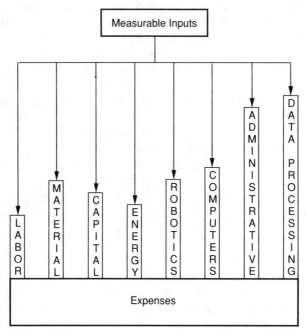

FIGURE 8.14 Input components considered in the task-oriented total productivity model.

Step 1. Project Introduction and Familiarization Sessions: In this step, the objective of the project was clearly defined and presented to all levels of management and employees. The manufacturing processes, procedures, cost, and types of printed circuit boards were explained. The key personnel in charge of various manufacturing activities were made to understand their role during the study.

Step 2. Total Process Analysis: The input/output of the robotics-aided task and the manual task was performed. The total process analysis provided the basis for balancing production parameters, quantifying the various input and output elements, and managing the productivity and quality of the robotics-aided task and the manual task.

Step 3. Allocating Overhead Expense: In this step, allocation criteria for overhead expenses were developed. Proportional contribution to the total number of insertions in printed circuit boards was used to allocate overhead expenses to the various input and output elements of the tasks.

Step 4. Base Period and Deflators Selection: The input and output elements were deflated to remove the effect of price changes. The deflator in this study was assumed to be 1.00 for all input and output components. The deflator did not vary because the short study period was the reference period to which all productivity values and indices were compared. The base period was assumed to be the first week of the study.

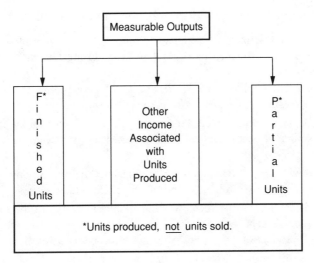

FIGURE 8.15 Output components considered in the task-oriented total productivity model.

Step 5. Data Collection Design: Forms were designed to capture all the input and output elements needed for TOTPMM and to track all the project activities. The forms were tailored, for the most part, to specific input, output, or activity.

Step 6. Personnel Training and Testing of Data Collection Instrument: The project participants were trained how to collect the input and output components
needed for TOTPMM. The data collection instruments were tested and revised for clarity.

Step 7. Data Collection Synthesis and Computations: The input and output component values collected periodically were synthesized and used to compute productivity values and indices.

Step 8. Trend Analysis and Interpretation of Results: The total and partial productivities for both the robotics-aided task and the manual task were analyzed periodically. The strategy was to understand the reasons/causes for the increases or decreases in productivity values and indices. Then specific actions to address the problems identified were implemented.

Step 9. Project Status Meetings: Meetings were held twice a week for one hour specifically on the status of the project. At this meeting all the project participants from various functional areas presented a summary of key issues, problems, and suggested solution strategies as they related to their specific tasks. These meetings were a good forum to discuss all potential implementation problems and solution strategies.

Step 10. Project Coordination: The project activities were coordinated by a project manager. The project manager provided the focal point for team work, progress reports, task scheduling, and time management.

8.6.6 The Project Manager's Role during Implementation

In addition to the coordination functions and responsibilities presented in Step 10, the project manager performed the following tasks during the implementation of the robotic device.

1. Designed and implemented PERT and CPM networks to track all project activities. A computer data base was used for this purpose.
2. Designed and implemented a resource allocation scheme for the project. The resource allocation scheme implemented was reviewed by all levels of management for approval.
3. Worked with project team members to manage the project's ongoing shifting priorities, technical problems, and other human-related issues.
4. Ensured that a balanced focus was maintained to carry out the project. Key issues such as cost, process definition, quality control, productivity levels, and total production objectives were reviewed periodically.
5. Acted as the project advocate in all business meetings.
6. Managed the project task across a matrix type of organization.

8.6.7 Results

As shown in Figures 8.16, 8.17, 8.18, and 8.19, when the manual method of assembling printed circuit boards was changed to the robotic device method, there was an increase in labor and total productivity. The robotic device method was repeatedly able to work faster than manual method. The programming sequence of the robotic device eliminated extra motions performed using the manual method. Excessive transportation, handling, grasping, and reaching motions were eliminated using the robotic device method. While the robotic device used a consistent

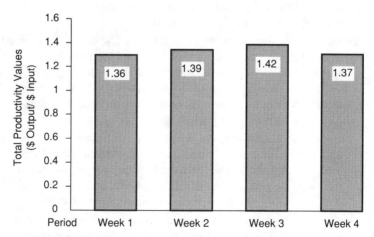

FIGURE 8.16 Total productivity values for the manual PCB task.

CASE STUDY SIX: JUSTIFYING THE USE OF ROBOTICS 175

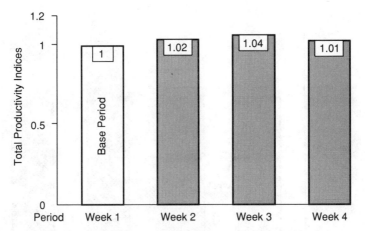

FIGURE 8.17 Total productivity indices for the manual PCB task.

pace, the manual method required operators to work at their own pace and also required more motions because of the way the work envelopes were designed. The robotic device method, however, was found to have increased energy, capital, robotic operating expense, data processing and other administrative expenses. More electricity was used to support the mechanical motion of the robot movements; the manual method used electricity only for the lightbox. The robotic device required more indirect effort supervision, more maintenance, safety procedures, programming support, and tooling costs. These requirements accounted for the decreases

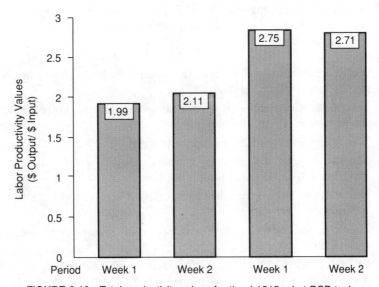

FIGURE 8.18 Total productivity values for the J-1515 robot PCB task.

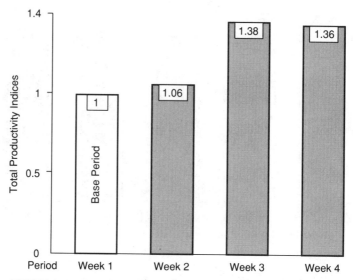

FIGURE 8.19 Total productivity indices for the J-1515 robot PCB task.

in most of the partial productivities (see Figure 8.18). However, the gains in labor productivity were large enough to offset the impact of the operating costs. It is important to improve the design of the robot to minimize the operating cost; otherwise the total productivity might be significantly affected in the long run.

8.6.8 Conclusions

Based on the successful implementation of the specified robotic device, the following 10 points are recommended when implementing a new technology in a production environment.

1. Obtain the total support of the total organization and vendors involved in the implementation project.
2. Set realistic implementation goals and objectives and maximize the effective use of resources.
3. Plan implementation activities ahead of time.
4. Maintain internal and external contacts with key people who have the expertise to help. Do not reinvent the wheel.
5. Find out who has implemented such a project before, and learn from their experiences.
6. Follow up and monitor project activities closely. Close open issues quickly and effectively.
7. Sell the benefits of the project effectively to the total organization. Show implementation progress to everyone involved.

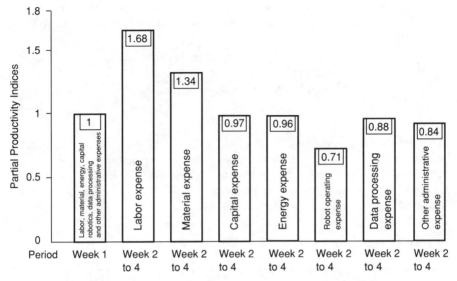

FIGURE 8.20 Partial productivity indices for the J-1515 robot PCB task.

8. Keep a positive team attitude toward implementation.
9. Motivate project participants to do their best and reward good accomplishments in a timely manner.
10. Encourage ongoing improvement, evaluation, and assessment of the project activities.

The successful implementation of a high-technology device in a production environment requires a close link between system designer and manufacturing personnel through the concept of early manufacturing involvement (EMI) in the design phase. The teamwork approach should be used to handle and manage all project activities. The project manager's coordination skills and technical talent are needed to ensure that all key issues relating to the project are handled successfully. Most projects fail not because they are unsound, but because the mechanisms for implementation were not carried out properly.

8.7 CASE STUDY SEVEN: PRODUCTIVITY MEASUREMENT IN A GROUP TECHNOLOGY (GT) PRODUCTION ENVIRONMENT

8.7.1 Total Productivity Measurement in a GT Environment

The technology-oriented total productivity measurement model (TOTPMM), presented in Section 4.7, is adapted for use in a GT production environment. The production environment is classified into uniform cells. Each cell has the property of being able to process a similar product mix, supported by the same system functions. For each production cell, the total productivity is measured as follows:

$$\text{Total productivity of GT cell } i \text{ at site } j \text{ in period } t = \frac{\text{Total measurable output of GT cell } i, \text{ at site } j, \text{ in period } t}{\text{Total measurable input of GT cell } i \text{ at site } j, \text{ in period } t}$$

The total factor productivity (TFP) of GT cell i is the ratio of total measurable output to the associated labor and capital input. Partial productivity (PP) of GT cell i is the ratio of total measurable output to one class of input, such as labor, material, energy, and capital. The possible tangible input and output components involved in a GT cell are specified in Section 4.7.

8.7.2 Machine and Part Grouping Techniques

One requirement of group technology applications is the grouping of machines into cells and parts into families. This process can be cumbersome. Possibilities for grouping N objects into M groups result in a Stirling number of the second kind; for example, the number of possible ways of grouping 25 machines into five cells is an astounding 2,436,684,974,110,751. Moreover, the complexity is greatly enhanced if the number of groups is unknown. Three possible classification schemes are

1. Form part families and then group machines into cells (part family grouping).
2. Form machine cells based on similarity in part routings and then allocate parts to cells (machine grouping).
3. Form part families and machine cells simultaneously (machine-part grouping).

The machine-part grouping classification scheme is most widely used. Two manual techniques of machine-part grouping are production flow analysis and component flow analysis. Both are more comprehensive in scope than algorithms for forming part families and machine groups.

8.7.3 Allocation Criteria for Overhead Expenses

To allocate overhead expenses to the various input components of TOTPMM, use of the complexity factor for each GT cell is recommended. This complexity factor is obtained from the functional relationships of five major variables shown in the following expressions. All variables are normalized to the same time period and ratio base:

$$\text{GT cell: Complexity Factor} = f(\text{Mir, Meu, Cpt, Mur, Dhu})$$

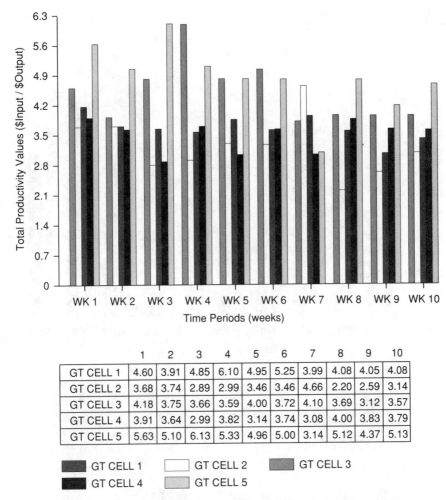

FIGURE 8.21 Total productivity values for GT cells 1,2,3,4, and 5.

where

  Mir = Machine insertion rate
  Meu = Machine energy utilization rate
  Cpt = Chip preparation time
  Mur = Machine utilization rate
  Dhu = Direct hours utilized

8.7.4 Study Results

The TOTPMM was applied in a GT production environment—a manufacturing plant that produces unit chips to customer orders. The chip-manufacturing process

180 CASE STUDIES IN TECHNOLOGY MANAGEMENT

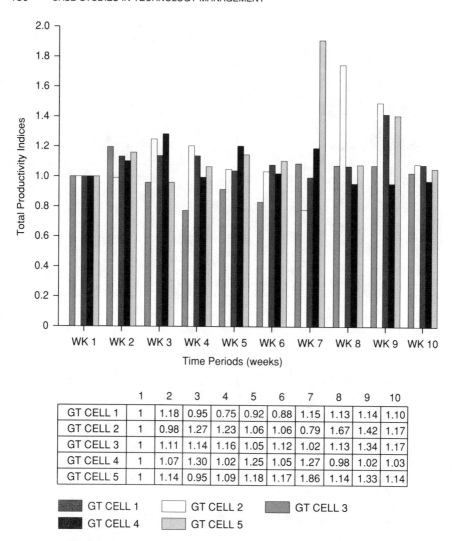

	1	2	3	4	5	6	7	8	9	10
GT CELL 1	1	1.18	0.95	0.75	0.92	0.88	1.15	1.13	1.14	1.10
GT CELL 2	1	0.98	1.27	1.23	1.06	1.06	0.79	1.67	1.42	1.17
GT CELL 3	1	1.11	1.14	1.16	1.05	1.12	1.02	1.13	1.34	1.17
GT CELL 4	1	1.07	1.30	1.02	1.25	1.05	1.27	0.98	1.02	1.03
GT CELL 5	1	1.14	0.95	1.09	1.18	1.17	1.86	1.14	1.33	1.14

FIGURE 8.22 Total productivity indices for GT cells 1,2,3,4, and 5.

was divided into five GT cells. Data for input and output components from each GT cell was collected for 10 equal time periods. The input and output values for each GT cell were recorded in terms of dollars. The values were not deflated because the study period of 10 weeks was not long enough to include any major significant variation in prices. However, the productivity values and indices were analyzed with reference to the same base period. The total productivity values and indices for all five GT cells are presented in Figures 8.21 and 8.22, respectively. To access the difference in productivity gain between a group technology application and a typical machine process layout application, comparisons were done between

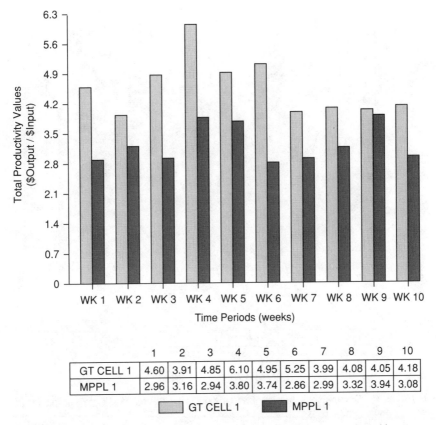

FIGURE 8.23 Comparison between total productivity values of GT cell 1 and machine process pattern layout 1.

the total productivity values and indices of GT cell 1 and a similar machine process pattern layout; they are shown in Figures 8.23 and 8.24, respectively.

In all five GT cells there were significant fluctuations in productivity values and indices during the first three weeks of the study. This was primarily due to start-up problems within each GT cell. It was observed that productivity increased in all GT cells during the later part of the study (weeks 4–10). These increases were obtained primarily from thorough characterization of all machines and tools and the understanding of the relationships between the input-output rate between cells, reduction in defect levels within each cell, and thorough balancing of all production parameters such as cycle time, work-in-process inventory, setup time, parts handling sequencing, and overall logistic controls.

When the productivity values and indices of GT cell 1 were compared to a similar machine process layout pattern, it was observed that the GT application had more increase in productivity gains. GT cell 1 was also found to reduce

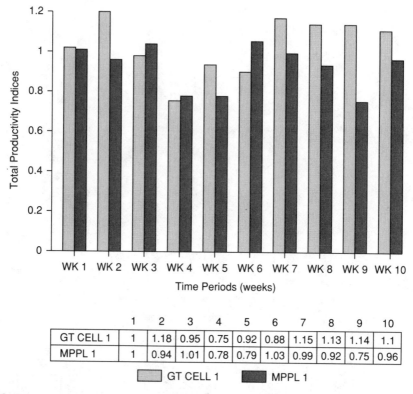

FIGURE 8.24 Comparisons between total productivity indices of GT cell 1 and machine process pattern layout 1.

product cycle time by 37% compared to the machine process pattern. Other benefits observed were reduction in work-in-process inventory and setup time for jobs.

8.7.5 Problems Encountered during Implementation

Initially, it was difficult to obtain TOTPMM input and output components from the existing data collection system. Modification was made to enable the input and output components for each GT cell to be collected. A computer data base and program was set up for TOTPMM computations. The GT cells' complexity factor was very difficult to obtain. After several iterations of machine variables, overhead expenses utilization, and other factors, the complexity factors were obtained through a close approximation technique. However, if the production process had been initially designed to have a streamlined pipeline for all input and output components, it would have been easy to obtain the GT complexity factor. Balancing the input and output between GT cells is not an easy task. Several simulation runs were performed before an optimal balanced line was obtained. The key to successfully balanced GT cells is thorough machine and tool characterization.

Detailed process characterization also enabled production parameters such as cycle time, setup time, and sequencing times to be controlled. A well-trained workforce that included a machine maintenance technician, operators, engineers, and a system analyst was necessary to resolving the technical issues that arose during the implementation. The additional cost to support the sophisticated GT cells at the beginning also contributed to the decline in productivity values and indices in some of the GT cells.

8.7.6 Conclusions

This is perhaps the first attempt to measure productivity in a group technology (GT) production environment. It was observed that the productivity values and indices provided the basis to identify areas within each GT cell that have potential for improvement. The productivity measurement within the GT environment also facilitates better resource planning, effective monitoring of all actions to improve performance, and comparison of performance among GT cells and in the firm as a whole. Although there are many techniques for machine and process characterization, additional tools and techniques are needed for machine and parts grouping for GT application, line balancing within GT cells, simulation of GT parameters, and understanding of the total complexity of GT applications.

GT application offers tremendous benefits if designed and implemented properly. It is more cost-effective than the traditional machine process layout pattern. However, both GT and the machine process layout pattern should be applied where they prove feasible. The technology-oriented productivity measurement model (TOTPMM) used in this study also has potential applications for technology justification and productivity measurement at the product, customer, task, and firm levels.

8.8 CASE STUDY EIGHT: IMPLEMENTING KNOWLEDGE-BASED EXPERT SYSTEMS

A knowledge-based expert system (KBES) was developed to assist the technician in performing defect isolation tasks. The KBES is used by the technician to select the appropriate types of symptoms and then the right type of action. The KBES was developed to provide troubleshooting assistance to technicians through an existing test system. The various relationships were all stated in the form of English-like rules and solution strategies.

In implementing knowledge-based expert systems, the most important ingredients for success are presented in the stages outlined in Figure 8.25. These implementation stages are suggested as a guide for potential users. Different KBES applications might require slight modification of the method of implementation. The KBES was developed by 10 domain experts with an average experience base of six and one-half years.

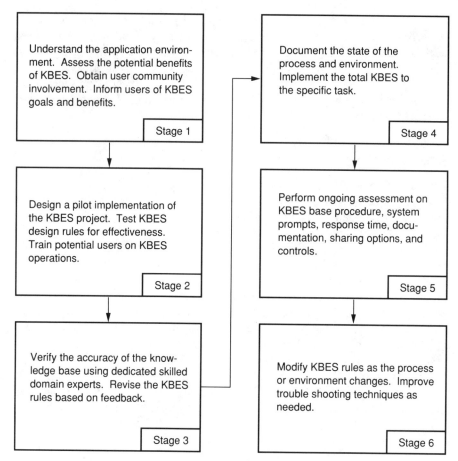

FIGURE 8.25 Implementation stages for knowledge-based expert systems.

8.8.1 Operation Sequence

The sequence of operation is as follows:

Step 1. Machine error occurs during test cycle.

Step 2. Technician initiates KBES consultation based on characteristics of specific failure.

Step 3. Inference engine processes KBES rules to determine the failure probable cause.

Step 4. Inference engine recommends cause(s) of action for probable failure cause.

Step 5. Technician provides additional failure symptoms through an interactive dialogue with the KBES program.

Step 6. Steps 2, 3, and 4 are repeated until complete resolution is obtained.

CASE STUDY EIGHT: IMPLEMENTING KNOWLEDGE-BASED EXPERT SYSTEMS

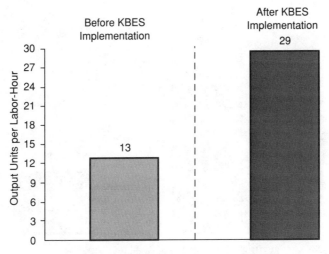

FIGURE 8.26 KBES impact on output per labor-hour.

8.8.2 Case Study Result

The results obtained from the implementation of KBES are presented in Figures 8.26, 8.27, 8.28, and 8.29. The output per labor-hour improved significantly after the implementation of KBES. There was also a significant reduction in the training for technicians and diagnostic time for failures. The average consultation time was also reduced.

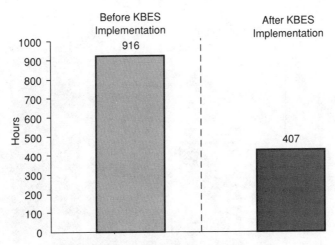

FIGURE 8.27 KBES impact on training time.

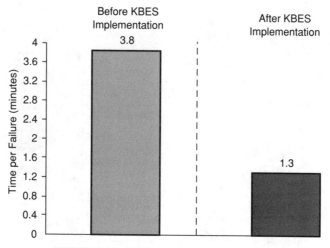

FIGURE 8.28 KBES impact on diagnostic time.

8.8.3 Conclusions

KBES can offer significant improvement in productivity when developed and implemented properly. The implementation strategy used for KBES should include total involvement of the developers and users in the design and implementation phases, a clear set of objectives, use of proper domain experts, selection of appropriate knowledge base and rules, construction and evaluation of KBES prototype, training of potential users, and ongoing assessment of the KBES application environment. KBES application could easily fail under the following conditions:

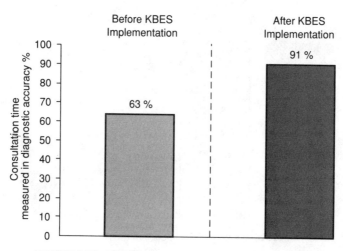

FIGURE 8.29 KBES impact on consultation response time.

- The problem environment changes.
- KBES makes the job more difficult.
- Real problem not completely understood.
- Real problem is too complex to handle.
- KBES does not provide complete solution.
- Users reject KBES application for unknown reasons.
- The system is not cost-effective and productive.

In the world of knowledge-based expert systems, the expensive, scarce resources will be domain experts, software and hardware for expanding a base, knowledge engineers, information asset security, and methodologies to maintain a stable process in both production and service work environments.

8.9 CASE STUDY NINE: A PRODUCTIVITY PERSPECTIVE TO MANAGE TECHNOLOGY CHANGES†

Peters (1987, p.57) rightly points out that managing discontinuities "will require us to specify and measure technological performance, both output as well as input, and to seek and understand alternative approaches and their limits." Since technology and productivity are closely related, it becomes a natural prescription to link the two. Clark, Hayes, and Lorenz (1985) also see this inevitable linkage. Therefore, we propose a productivity methodology that evaluates rationally the appropriateness of adopting a new technology or abandoning an existing one.

The CTPM generates a Comprehensive Total Productivity Index (CTPI) for a given system for any time period, with reference to a base or reference period. The CTPI is given by the following expression:

$$\text{CTPI} = \text{TPI} \times \text{IFI}$$

where:

TPI = Total productivity index
IFI = Intangible factor indexes

The TPI reflects the percent change in total productivity value (a ratio of total tangible output to total tangible input) from the base period. The tangible outputs and inputs are inherently measurable. For example, total tangible output is the sum of finished and partially finished units, plus other outputs such a dividends from securities, interest from bonds, and other income. The total tangible input includes all the inputs consumed in a given period to produce the above-mentioned tangible outputs. There are five such input factors: human, material, capital (fixed and working), energy, and other expense inputs.

The IFI captures the percent change in the intangible factor value (IFV) for the system from the base period. This intangible factor value is computed from a set of

---

†This case study was prepared by Dr. David J. Sumanth, professor and director of Productivity Research Group, University of Miami, Coral Gables, FL. The material is included with permission.

TABLE 8.3 Typical Intangible Factors Considered in the CTPM

I. Customer-related factors
- Product quality
- Product reliability
- Service quality
- Price competitiveness
- Product/service loyalty

II. Market-related factors
- Market standing
- Company/organization image

III. Society-related factors
- Community attitude
- Pollution effect

IV. Process-related factors
- Process timeliness
- Process time
- Process effectiveness
- Process efficiency

V. Employee-related factors
- Job satisfaction or employee morale
- Salary or wage raises
- Productivity gainsharing
- Job security
- Employee loyalty
- Employee turnover

VI. Vendor-related factors
- Stockholder's/owner's financial benefit
- Stockholder loyalty

Note: Each company/enterprise can define its own intangible factors apart from or substituting these, depending upon its particular needs.

user-defined intangible factors. The number of such factors can be as large as the user finds necessary, but they are classified into seven *basic* categories: customer related, market related, society related, process related, employee related, vendor related, and owner related (see Table 8.3). Each of the intangible factors is assessed by a numerical rank based on an appropriate scale that is agreed upon by experts (three methods of weighting are generally feasible), and a composite score is obtained as the IFV for any particular time period t (base or reference period is $t = 0$). The ratio of the IFV in period t versus $t = 0$ gives us the IFI. Once the TPI and IFI are computed, the CTPI is determined.

8.9.1 Application of CTPM to Manage Technology Discontinuities

When a new technology is a candidate to replace an existing one, the basic strategy proposed here is to ensure that:

$$\text{CTPI}_{t+T} > \text{CTPI}_t \text{ for } T = 1, 2, \ldots, n$$

where n represents the time periods (See Figures 8.30 and 8.31).

Since the CTPM takes into consideration all the "user-definable" factors of output and input, and since profitability is directly related to total productivity, the previous criterion for choosing a technology when a discontinuity takes place is comprehensive and practical from the point of view of "good business practice." It balances the technical, economic, social, cultural, and even political factors affecting an enterprise when major changes occur in product technologies or process technologies.

8.9.2 Example

Consider a leading manufacturer of personal computers. Until now, the company's personal computer had Type A microprocessor technology. The company now wants to introduce a Type B microprocessor technology with much higher CPU speeds and multitasking and multi-user capabilities, so this company is planning a new product technology. As a result of this new technology, some of the production processes would change drastically. Furthermore, the marketing and advertising need to be refocused. The company wants to discontinue the old product (with Type A processor) totally within a few months and offer only service to the old customers. Clearly, this situation poses a strategic challenge to the company's top management. The company builds a prototype and tests on a selected basis.

At this point, most traditionally managed companies would produce and market the product aggressively. However, if the company were to rationally approach

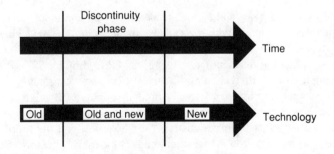

FIGURE 8.30 Schematic representation of the "technology discontinuity phase" for the purpose of application of the CTPM. (Source: Sumanth, 1987).

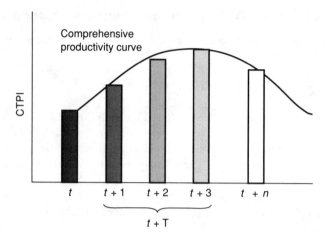

FIGURE 8.31 The comprehensive productivity curve (Source: Sumanth, 1987).

managing the discontinuity (from the Type A to Type B processor technology), it would apply the CTPM. Suppose, from the data collected, it obtains the following values:

$$TPI = 0.920$$
$$IFI = 1.225$$

Then

$$CTPI = (TPI) \times (IFI) = (0.920) \times (1.225) = 1.127.$$

This means that the introduction of the Type B processor technology will initially reduce the total productivity level of the company by (1 - 0.92) = 0.08, or 8%, because of heavy capital expenditures to change the production processes at the company's plants and the consequent high fixed capital input. Now the total productivity is given by $TP = O/(I_h + I_m + I_c + I_e + I_x)$ where O is the tangible output, and I_h, I_m, I_c, I_e, and I_x are, respectively, human input, material input, capital input, energy input, and other expense input. With the new technology, the company's I_c will rise drastically and, as a result, the total input is expected to increase faster than the total output, causing the TP to drop by 8%.

On the other hand, with technology B, the IFI is expected to increase by 22.5% because of sizable improvements in (1) product quality, (2) product reliability, (3) customer satisfaction, and (4) stockholder satisfaction. These first three are certainly most important from the customer's viewpoint, and the fourth one is of importance to company's top management and the board of directors. Considering that the personal computer market is highly competitive now, with "clones" from Japan, Taiwan, and other countries, the company correctly observes that the 8% drop in total productivity due to Type B technology is well compensated by the 22.5% gain in the IFI. From the break-even point in total productivity, the company is expected to operate at a comfortable total productivity level. Therefore, the company decides to go for the new Type B technology and aggressively market the product in order to generate more output and higher total productivity.

8.9.3 Advantages of the CTPM Perspective

The use of a productivity perspective, with the comprehensive total productivity model (CTPM), to manage technology changes has a number of strengths:

1. *SYSTEMS VIEW.* All real-world situations are systemic in nature and the CTPM strongly reflects this view. It considers the joint and simultaneous impact of *all* resources whenever new technologies are introduced. The CTPM's strength lies in the fact that quality, reliability, effectiveness, efficiency, marketability, and all such performance measures are captured in an integrated manner, thereby recognizing their interactive nature.
2. *TANGIBLE AND INTANGIBLE FACTORS.* Both the tangible (easily quantifiable) and the intangible (traditionally unquantified) factors of technology impact are considered in the CTPM. For example, the interest of customers, employees, vendors, stockholders, and society are all factored in, thus minimizing the risks of customer alienation, employee dissatisfaction, stockholders' unhappiness, and the like.
3. *RELATIONSHIP TO PROFITABILITY.* Because the CTPM has the total productivity component, which is directly related to "productivity-oriented profit," the CTPM can easily be understood and appreciated by the management. A company's management can easily pinpoint the contribution of the newly acquired technology to the profits.
4. *INTEGRATABILITY.* The CTPM framework forms a core philosophy around which business, technology, manufacturing, and marketing strategies can be woven, thereby providing a unified direction at all levels of management in an enterprise.
5. *ABILITY TO ADAPT OTHER TECHNIQUES.* Quantitative models such as simulation, mathematical programming, and statistical process control can be applied to monitor the adaptation of new technologies. The CTPM provides a natural and intuitively appealing mechanism to do "what-if" analyses using such techniques.
6. *DATA COLLECTIBILITY.* Data can be collected at different levels of interest: corporate (firm), plant/division, product/service unit, department, work unit, or task. Thus, the impact of new technologies can be assessed at several levels of interest to top management, middle management, and operational management.
7. *LINKAGE TO GAIN-SHARING SYSTEMS.* Employee incentives and gain-sharing systems can be tied to the CTPM rather than the traditional, suboptimal bases. Teamwork becomes more natural in the CTPM environment than in the traditional, labor-productivity driven settings.

8. *MANAGEMENT OF COMPETITIVENESS*. Instead of reacting suddenly to external environmental factors of business, the CTPM offers a managerial tool to more broadly understand the role of technology and to evolve competitive strategies in a more holistic, but systematic manner.

The CTPM goes beyond being just a tool. It offers a different philosophy of management that provides an objective perspective to evaluate any alternative, not just technology, when decisions must be made. In today's corporations where being the largest is not necessarily being the best, there must be refreshing approaches to evaluate alternatives at all levels of an enterprise. This methodology, though in abbreviated form for the sake of space, offers such an approach in minimizing the dangers of being caught unprepared in technology discontinuities, which are not only there to stay, but are going to occur with greater frequency in the years to come. As Horwitch (1986) rightly points out, we must have a strategic perspective even in a rapidly changing, technologically complex world. In fact, the management of technology (MOT) is going to be an important field in the years to come, and the seeds for such thinking are already being sown (National Research Council, 1987). New approaches must also be devised to assess the performance of managers who would be increasingly associated with the management of technology in general and discontinuities in particular (Roberts and Fusfield, 1981; and Dearden, 1987).

ACKNOWLEDGMENTS

The author wishes to thank the authors of the Case Studies one, two, three, and nine for their contributions to this chapter. Each contributor is entitled to full technical credit for the study provided. All the contributions were included in this chapter with permission. Case Studies four, five, six, and seven were funded by grants from IBM and the Social Science Research Council (SSRC), U.S. Department of Labor. The opinions expressed in the studies do not represent the views of SSRC or IBM; the author is solely responsible for the contents.

9
CASE STUDIES ON SYSTEMS INTEGRATORS AND COMPUTER-INTEGRATED MANUFACTURING SYSTEMS

This chapter presents real-life case studies that describe the role of systems integrators and the adoption of Computer Integrated Manufacturing Systems (CIMS). Although the recommendations, methodologies, and procedures offered pertain to CIMS projects, most of the case study recommendations can also be applied to the implementation of other high-technology systems with minor modifications.

9.1 INTRODUCTION AND DESCRIPTION OF THE ENVIRONMENT

Since 1975, over 5 million U.S. manufacturing jobs have been lost. These losses have been greatest among the hundreds of thousands of small manufacturing firms with fewer than 100 employees, who conduct over 75% of America's machining, repairing, and refitting services. Service manufacturers perform vital repair and retrofitting services, and provide critical spare parts inventories to the military, consumer durables, and industrial machinery sectors. The only means of survival for most of America's service manufacturing firms may be their adoption of computerized integrated manufacturing systems (CIMS) that will significantly improve their product quality and throughput.

It is paradoxical that new manufacturing technologies and computerized systems abound, yet few service manufacturers are highly automated. The many explanations for this paucity of CIMS applications include ignorance of the issues, inability of the small firms to articulate their problems, a lack of vendors who are willing to work with small firms, inability of small firms to design their own systems, the small firm's lack of understanding of how to integrate personnel and equipment, the lack of a systems perspective among the small firms, inability or unwillingness of small firms to make commitments to untested systems, and lack of technical familiarity with computer networking. Archaic organizational structures that inhibit interdepartmental collaboration and reward systems that foster information hoarding are common phenomena within small firms that raise major barriers to CIMS. Institutionalized capital budgeting procedures are another major barrier. Traditional capital budgeting procedures ignore the benefits of flexibility, quality improvements, shorter lead times, customer responsiveness, and design to customer needs that make the difference between success and failure. Thus, a need has arisen for specialists who are interested in and can help small service manufacturers select and design CIMS for their particular applications.

9.1.1 Emergence of the Systems Integrator

In response to this need, a new form of management information service purveyor is now emerging: the systems integrator (SI). SIs are commercial change agents who sell client-specific engineered information systems that aid management decision making. Though many SIs are information equipment manufacturers and vendors, the SI's role is to link vendors to users. The SI's loyalty is to the client, serving as information consultant and technical counselor to help develop and implement systems that are optimal for that client, no matter who manufactures those systems.

SIs are specialized by markets and applications. For example, MSIs (manufacturing systems integrators) specialize in tailor making factory management systems that integrate product planning, CAD, computer-aided engineering (CAE), CAM, inventory management, order-billing, customer service, and real-time financial control functions. Each system is tailored to the individual needs of a particular plant. Similarly, CSIs (computer systems integrators), ESIs (electronic systems integrators), and DSIs (delivery systems integrators) develop highly specialized systems for their respective industries. The focus of this study is on MSIs.

9.1.2 Role of the MSI

The unique features provided by the MSI are vendor independence, client-specific responses, and integrated systemwide solutions. Because MSIs are independent agents, an MSI can select the best product or service for the individual client, irrespective of the manufacturer or vendor. Because the MSI's future depends entirely on success in fulfilling client needs, MSIs are responsive to individual differences in wants and needs. The MSI's work is based on the system philosophy: Systems problems are only solved through total integrated solutions. For example, under this philosophy, factory automation would not be viewed as a matter of simply installing the best available CAD or CAM component. Rather, the systems philosophy would examine the entire system of components and redesign or fit them together as required to create a balanced system. This requires major client commitments to entirely new technologies and their associated new ways of doing business.

The number of MSIs in the United States is increasing at a very rapid rate. For example, in 1983 there were only a few MSIs, but by 1988 there were over

250. Moreover, a large variety of types of MSIs is emerging, ranging from the independent consultant to the large MSI departments of major computer manufacturers, the MSI operations of large accounting houses, and various consortia of MSIs. Many MSIs are entrepreneurial efforts within conglomerate firms that, for various reasons and with varying success rates, have entered the growing market.

9.2 PROBLEM FOCUS AND METHODOLOGY

America's service manufacturers are in deep trouble. Because most of America's military, consumer, and industrial repairs and equipment maintenance are done by service manufacturers, the health of the industry is essential to the economic and political welfare of the United States. State and federal aid are available, but will not begin to solve the industry's problems.

The MSI institution and its infrastructure may be the industry's greatest hope. However, because of their relative newness, very little is known about MSIs, their modus operandi, and their possible impacts. Potentially, MSIs and CIMS could play a significant role in reversing America's declining manufacturing prominence. The adoption of CIMS will surely have dramatic impacts on the culture and operation of service manufacturing firms, who have traditionally functioned in a handicraft fashion. Much more research is needed to fully understand the role of the MSI and to develop detailed prescriptions for effective MSI activities.

This exploratory study chronicles the activities of four MSIs, in four successful cases and four unsuccessful cases of CIMS adoption within the service manufacturing industry. The purpose of the study was to learn more about the operations and the effectiveness of MSIs as change agents in encouraging the adoption of computerized manufacturing systems (CIMS) among manufacturing service firms. Because of the exploratory nature of this study, the focus was on generating hypotheses and propositions through detailed case studies. The objective was to elucidate emerging phenomena, as a basis for subsequent multi-site field studies.

9.2.1 Sample of Entrepreneurial MSIs

Four entrepreneurial, conglomerate MSIs were selected for study. Each MSI was an entrepreneurial effort within a larger firm, each offered a range of CIMS technologies and services, each was dedicated to assisting manufacturers, each had a large staff of professional personnel, and each had system development and marketing functions within itself. Entrepreneurial MSIs were chosen for study because they appeared to offer the greatest latitudes for learning about the MSI function and the MSI change agent process.

Each MSI was asked to reflect on its recent history of work with small service manufacturers, and to nominate one of its successful cases of CIMS adoption and one of its failure cases for study. These nominations were thoroughly discussed with the personnel at each MSI to ensure that the particular cases provided complete details and rich information. The sampling design is shown in Table 9.1.

9.2.2 Data Collection Methods

Using established instruments, archival methods, and interview protocols (Soulder, 1987, pp. 11–21, 23–47, 179–198, 216– 237), the following information was collected at each MSI: history; organization; long-range plans, goals, and missions; modus operandi; sales and promotional materials; line of services and products; product line mission statements; and other information that characterized and distinguished the MSI.

At each CIMS user (service manufacturer), information as collected on the history of the firm; its previous involvement with automated manufacturing technologies; its financial position; the set of barriers and promoters existing at the firm before and after the CIMS implementation; the perceived degree of success of the implementation; and the impacts of the implementation on the barriers, promoters, and financial health of the firm. For each of the eight cases studied here, the history of who did what and how the barriers to adoption either were or were not diminished were sought, and the ways in which the CIMS was designed and fitted to the user firm were detailed.

Extensive information was sought about why the CIMS application succeeded or failed, in terms of its impact on the following variables: throughput, production times, product quality, profits, costs, rejects, outputs, and employment. Success and failure were not measured here as dichotomous, mutually exclusive events; partial success along several dimensions is a more typical outcome than complete success or failure on every performance dimension. Thus, for the purposes of this study, the user was questioned about the performance of the CIMS and asked to score each of the above eight dimensions as either a "success" or a "failure." The user was then asked to provide an overall success or failure assessment of the CIMS.

As it turned out, when six or more of the above dimensions were scored "success," the users consistently assessed the CIMS as an overall success. When four or fewer of the above dimensions were scored "success," the users consistently assessed the CIMS as a failure. No other situations were encountered in the study.

Several telephone and face-to-face interviews were carried out with automated equipment vendors, industry consultants, other users, and other personnel within

TABLE 9.1 Sampling Design

	Adoption Cases: Service Manufacturer Code Name[a]	
MSI Site Code Name	Success Cases	Failure Cases
Alpha	Amanda	Stella
Beta	Barbara	Peggy
Gamma	Tilly	Charlene
Omega	Dolly	Arlene

[a]Each of these eight cases occurred at a different service manufacturer, as indicated by the eight different names.

the manufacturing service industry. The purpose of these interviews was to become better acquainted with the industry, its culture, and its challenges. Some of these interviews were also conducted to verify events and impressions gained from the interviews with the MSIs and CIMS users in the case studies.

9.3 RESULTS

9.3.1 Successful Cases

Amanda. Amanda was a very old, unprofitable facility in an economically depressed geographical region. Amanda's vice president of engineering spearheaded a "last resort" internal study to completely automate the plant. After lengthy deliberations, a modified automation plan was accepted by Amanda's top management, and MSI Alpha was chosen for the job on the basis of its low cost bid. Although it appeared many times that top management would cancel the automation project, it was completed and the plant successfully returned to profitable operations.

Amanda's labor force selection process was an important factor in the success of this case. All the old employees were laid off, the plant was completely reorganized, and entirely new job classes were established. A new labor force was hired on the basis of scores on a battery of psychological and performance tests. Since Amanda is one of the few large employers left in this geographic region, the labor force is grateful to be employed. They take enormous pride in their work, demonstrate a strong *esprit de corps*, and are fiercely dedicated to the success of the new facility. The new labor force participated fully in the CIMS design and implementation. They made many suggestions that improved the CIMS design. A unique management-employee team spirit that still endures was formed during the CIMS implementation.

Barbara. Barbara was a relatively new and financially healthy facility. Automation was an integral part of Barbara's long-range plan for maintaining future competitiveness. As in most CIMS applications, interfacing the varied computers and equipment that existed throughout Barbara was a major technical hurdle. The office computer was an IBM mainframe, the plant used several other different computers, and management wanted to interface everything through their IBM mainframe. There were many different equipment manufacturers represented in the plant, and none of them wanted anything to do with the interfaces.

The absence of skilled in-house factory engineering personnel at Barbara was another major hurdle. Like most U.S. plants, Barbara had eliminated its factory engineers and plant software designers in a cost-cutting move. This presented an implementation problem for MSI Beta, since there was no obvious in-house group to work with in the development and transfer of the completed CIMS. Success appeared to be enhanced because Barbara's management gave Beta a free hand in forming implementation teams. Beta was able to draw on any Barbara personnel it needed for various team assignments. Barbara's organizational culture was characterized by open interdepartmental communication, collaboration, and sharing of technical details. This provided Beta with the opportunity to work across departmental lines, to build many alliances, and to quickly resolve any technical issues.

Tilly. Tilly hired Gamma to evaluate the prospects of expanding and automating Tilly's old plant. Tilly's plant had been profitable, but faced the prospects of future losses as the cost of operating its old facility rose. Tilly's strategic business plans consisted of a series of phased improvements to the plant. Gamma concluded that nothing could be done: The plant was simply not well configured and had incorrigible logistics. A new CIMS plant was estimated to cost $500,000,000, and Tilly's management declined to make such a large investment in a lump sum. Tilly's entire management team consisted of sons, daughters, grandsons, and sons-in-law of the founder. They traditionally made decisions by consensus, were generally averse to risk and had not adopted a major innovation in over 50 years.

However, four factors combined to change management's decision. First, Tilly was a proud, publicly traded but family owned and controlled firm. It had a reputation as the best quality U.S. maker of a specialty product, and the family was devoted to maintaining this reputation. Second, because the firm was family controlled, Tilly's managers could afford to look a long way into the future without answering to an annual profit cycle or to the constancy of quarterly dividends; they could sustain a few loss years and not be deposed. Third, when Tilly management chanced to discuss their production process with a Japanese engineer, they were stunned to learn that the Japanese knew more about their process than they did. Fourth, Gamma's engineers demonstrated that Tilly's control and information systems were so archaic that Tilly sometimes unwittingly sold its product below cost. These last two events energized Tilly's engineering vice president to mount a successful campaign to convince the other managers to approve the construction of a new CIMS plant.

Several factors appear to be important to the success of this case. The real-time cost accounting system that Gamma installed early in the system development work demonstrated the benefits of better management control. Tilly's practice of hiring young college graduates and offering them lifetime employment provided a continuity of personnel, thereby promoting the long-term use of Gamma's CIMS once it was accepted. Two of Tilly's plant foremen refused to accept any software or hardware whose reliability and operability had not been thoroughly demonstrated by Gamma. Although this intransigence often caused delays in the development, it also ensured the quality of the product and led to a commitment from the foremen once the products met their standards. Tilly's consensus decision-making style ensured a companywide commitment once the decision to adopt the CIMS was made. The fact that Tilly did not have experienced personnel for the development and implementation taskforces proved to be an advantage: The personnel came to their assignments with open minds and strong willingness to listen to Gamma's suggestions. Because Tilly had no prior experience with CIMS, it accepted Gamma's suggestion that it loan key production personnel to several vendors under a unique internship agreement. Under this agreement, Tilly paid its employees' salary and the vendors agreed to involve the employees in their operations, as if they were the vendor's own employees. This policy proved to be an excellent hands-on training experience for Tilly's employees.

The counterpart Tilly and Gamma development teams appeared to be another key success factor. All the necessary functions— engineering, production, training, cost, software, and process— were represented on each team. All teams reported to their respective managements at Tilly and Gamma. The team members were intentionally drawn from all over their respective firms, Gamma and Tilly, and insulated from their usual daily activities so they could devote all their time to the work of developing and implementing the CIMS. The success of the counterpart teams appears to explain the excellent technology transfer that occurred between Tilly and Gamma.

Dolly. Although Dolly was a relatively new and profitable facility, competition was rapidly increasing and Dolly's market share was eroding. Dolly, whose public image was "big, dumb and rich," was generally slow to react to competitive and technological change. It was MSI Omega that first suggested and persuaded Dolly to accept a CIMS. Dolly's slow and deliberate decision-making style was a major barrier to Omega. Dolly's management listened attentively, but provided no feedback or direction. Omega personnel complained that they never could tell whether Dolly management liked the work or not. Once Omega discovered that Dolly wanted to be led, Omega became much more assertive and dogmatic about the designs, and the project swiftly moved to a successful conclusion.

9.3.2 Failure Cases

Stella. Although Stella was profitable, its management felt the need to automate to remain competitive. However, management feared making a wrong decision. Stella's vice president of engineering finally took the initiative, driving Stella to hire Alpha to design a CIMS. Stella's engineering department wanted a turnkey operation with a very simple CIMS. This particular CIMS fit the Stella culture, but did not provide the potency that Stella needed to achieve a long-term competitive advantage. Stella's ignorance of control systems was a major barrier. So was Stella's lack of tolerance for downtime and the risk of failure. To abate this fear, Alpha incorporated many back-up systems and redundancies into the design. This raised the system costs well beyond the original budget, and Stella's management canceled the work.

Peggy. Peggy was an old, unprofitable facility, and management saw CIMS as a road to recovery. Like Stella, Peggy desired a turnkey operation. It hired Beta with the understanding that Beta would design the system, hand it off to them, and then go away. However, Peggy's demands for turnkey operations appeared to severely limit the success of this effort. Peggy's management had little experience as a prime contractor. Some subcontractors were permitted to severely overrun their budgets, and none were carefully monitored for their technical progress. Peggy failed to monitor the work packages in proportion to their importance, with some trivial work packages receiving most of Peggy's attention. No incentives were provided to encourage the subcontractors to collaborate. Since Peggy had accepted only the lowest bids, the subcontractors' profits were so low that they used the cheapest equipment and contributed minimal effort. None of the contractors had a warm feeling for Peggy, and the entire atmosphere was one of mistrust, antagonism, and maximization of self-interest.

Peggy's obsession with cost containment would not allow any scope changes. This severely constrained the work, which naturally had to flex to meet changing events over the life of the project. Near the mid-point of the development, Peggy's management was driven to terminate the more exotic software developments as part of a cost-cutting mode. That reduced the capability of the product, which further aggravated Peggy's management and inclined them to react by withdrawing still more funds from the project! The end product was a substandard CIMS with minimal capabilities and features that has proven unsatisfactory to Peggy.

Charlene. Charlene was an old, unprofitable facility. Like Peggy and Stella, Charlene also desired turnkey operations. Charlene had a reputation as an autocratic employer, never involving any employees in new equipment or process implementations. Charlene's management felt that the employees should be able to use whatever was placed before them. High employee turnover and frequent strikes plagued the CIMS development, and the resulting CIMS performed far below everyone's expectations.

Arlene. Arlene was a relatively new and profitable facility, but management was inclined to automate to achieve the consistent quality that was now being mandated by many of their customers. The CIMS development was troubled from the start by Arlene's lack of appreciation for CIMS technologies. Omega complained to Arlene's top management that the middle managers were indecisive and failed to make design choices in a timely fashion. Arlene's top management took this as evidence that the firm was not ready to accept the sophistication of a CIMS. As a result, the original system was severely scaled back and a much less sophisticated and much less effective system was implemented.

9.4 CASE ANALYSES AND DISCUSSION OF RESULTS

These eight cases provide many lessons about the best ways to promote the adoption of CIMS and about the optimum behavior of MSIs. Some of these lessons reinforce the results from other studies reported in the literature and fortify the conventional wisdom. However, because of the exploratory nature of this study and the small sample sizes, these lessons should be treated as hypotheses and propositions for further testing.

9.4.1 Conditions Found to Promote Adoption

Seven conditions, or *promoters,* were found in the cases that appeared to promote the successful adoption of CIMS. These seven promoters are listed in Table 9.2 and discussed below.

Linked Champions. The Amanda, Stella, and Tilly cases demonstrate that inside-the-firm engineering groups can be prime movers for the adoption of CIMS, but only if they enjoy high status linkages to upper management champions and to plant floor–level champions. Because inside-the-firm engineering departments often play technology watchdog and gatekeeping roles for the firm, they are in a position to favorably influence the adoption of CIMS. However, it is not sufficient for engineering to be the only champion. There must be at least one higher-level management champion, and there must also be a champion at the floor level.

The Tilly case provides a vivid example of the importance of a wily, old-time, no-nonsense plant floor staff who accepted nothing less than complete operating success. The staff refused to accept anything that could not be demonstrated to repeatedly function well. Thus, natural leaders at the plant floor level appear to be a necessary ingredient in any formula for CIMS adoption. There are key low-level decision makers in any organization whose support, once won, can become an important promotive factor. For example, labor force participation was demonstrated as a potent tool for implementation at both Armanda and Tilly. When the labor force was excluded (Charlene) or simply neglected (Stella and Peggy), the CIMS efforts were much less successful.

Economic Strength of the Adopter. The experiences at Barbara, Dolly, Peggy, Charlene, and Arlene might suggest that new plants are easier to automate. Old cultures are more likely to resist new things, and supervisors who are enculturated in these older cultures may also resist new ideas. New plants may be easier to automate because management always wants the best for their new plants. When the economic viability of the plant becomes so bad that the adoption of CIMS becomes imperative for survival, the fear of failure and the associated regrets may also become so great that they effectively prevent adoption when the plant needs it most. The fear of adoption may thus increase as

TABLE 9.2 Seven Conditions That Promoted the Adoption of CIMS

Linked champions
Economic strength of the adopter
Family ownership
Japanese-style employment
Vendor internships for user employees
Well-developed user strategic business plans
Rigorous personnel selection

the plant becomes increasingly older, less competitive, and financially disadvantaged. The fear of new things may be least when profits are high and greatest when profits are low. Thus, the lesson seems to be: Target plants for upgrades when profits are high.

However, this situation is undoubtedly much more complex than the above analysis suggests. The adoption decision will be a tug of war among the relative strengths of the need to survive, the perceived uncertainty of a do-nothing policy, the felt risk of investing in an uncertain new technology, and the desire to simply close the plant and walk away. Interpersonal dynamics and political powers can tilt the balance either way. This area needs much more research.

Family Ownership. The Tilly case demonstrates the potency of the combination of family ownership and the heritage of a quality specialty product. Family pride in a quality reputation is a valuable asset, worth protecting even at considerable cost. Family ownership gives a firm the latitude to look far into the future, since the firm is not compelled to answer to an annual profit cycle or quarterly dividends. Such firms' management can take a few loss years without being deposed.

This author's bias was that family ownership would be the kiss of death for new technologies, condemning the firm to archaic practices forever. Indeed, Tilly's family ownership and consensus decision-making processes delayed its CIMS adoption. When the entire management team consists of a cohesive family of sons, daughters, grandsons and sons-in-law of the firm's founder, there is a sentiment that no decisions will be taken unilaterally. Therefore, the decision to adopt will be long, time-consuming and very circumspect. But once that decision is made, it may be very permanent.

Japanese-Style Employment. There is a prevailing notion that employment stability encourages adoption. The Tilly case demonstrates that firms with lifetime employment policies, that is employment from college to retirement, are initially reluctant but thereafter committed adopters of CIMS. Lifetime employment is the Japanese style of employment (Cohen and Zysman, 1987). The two production foremen who insisted that Gamma's CIMS meet stringent capability requirements were lifetime Tilly employees. They so strongly identified with the company that they felt a personal responsibility to ensure that only the very best technology was being implemented at Tilly. Once they were convinced of the viability of Gamma's CIMS, they became strong champions of it.

Vendor Internships for User Employees. The unique mechanism of free loaning of one's own personnel to vendors is an effective way to aid in the transfer and training for CIMS. The experience at Tilly testifies to the potency of this technique. Tilly's employees received hands-on training under the close supervision of the technology originators. They were privileged to experience the technology first-hand, in use, in plants comparable to Tilly's. This brought them into contact with other users, affording them the opportunity to exchange perceptions and insights with peers in other organizations. Peer-to-peer interactions with other adopters outside the firm have been shown to correlate with the rapid diffusion of innovations (Rogers, 1983, pp. 25–78, 109–127, 189–226). So has hands-on training (Souder, 1987). The vendor internship program has both of these characteristics.

Well-Developed User Strategic Plans. Research has demonstrated that it is important for the user to have a well-developed strategic business plan as a basis for the adoption of CIMS (Cohen and Zysman, 1987). The Barbara and Tilly cases demonstrate that when the client has well-defined wants, well-defined work goals, and a well-defined time horizon, the CIMS design and adoption will proceed much more smoothly. The Stella, Peggy, and Charlene cases show how the lack of good planning for CIMS can in fact disrupt its successful development and implementation. The client's corporate plan and strategic market analysis are key success factors. If they are well developed, the client and the MSI can assemble a joint team and conduct a complete needs analysis, a thorough technology survey, and a systematic equipment selection, then make accurate budget projections. For effective CIMS adoption, the adopter firm must be looking well beyond the current one-year horizon. The fact that few firms have well-developed long-range plans has been one of the big barriers to CIMS implementation in America (Cohen and Zysman, 1987).

The systems integrator's heavy emphasis on front-end planning will be resisted by the user who is proactively oriented to early completion of the implementation. This was demonstrated by the unsuccessful Peggy and Charlene cases. Both were unwilling to pay for the front-end planning and design work planning needed to develop a system tailored to them.

Rigorous Personnel Selection. Studies have demonstrated the importance of careful personnel selection in staffing CIMS plants (Cohen and Zysman, 1987; Rogers, 1983). This author has been impressed with the rigorous selection and training of workers and their total involvement in the start-ups at Amanda and Tilly. These practices generated a feeling of "the company is providing a place for us and we should not let them down," along with a tremendous team pride and *esprit de corps*. Hand picking of employees through various psychological and performance tests to staff the new CIMS appears to provide the most effective personnel force.

9.4.2 Conditions Found to Retard Adoption

Sixteen conditions, or *retarders,* were found that appeared to retard the adoption of CIMS. These sixteen retarders are listed in Table 9.3 and discussed below. It may be noted that the number of retarders exceeds the number of promoters discussed above, reflecting the immature state of the art in the adoption of CIMS. In other studies of this nature, it has been found that as the state of art in adoption advances, the number of retarders decreases relative to the number of promoters (Souder, 1987).

TABLE 9.3 Sixteen Conditions That Retarded Adoption of CIMS

Fears about maintainability
Fears about reliability
Low tolerance for critical risks
The too simple/too complex dilemma
Islands of technology
Internally contradictory user goals
Lack of experience in prime contracting
A nonparticipating client
Desire for turnkey technologies
Lack of in-house manufacturing engineering
Client over-concern for controlling CIMS costs
Tenacious owners
Unrealistic expectations about CIMS
Paralysis by fear
Internal status differentials
Internal organization climates

Fears about Maintainability. The user's fear of maintainability was a barrier in all eight cases. None of the users welcomed CIMS with completely open arms. Even the sites that desired turnkey operations (Stella, Peggy, and Charlene) were apprehensive about maintenance. Highly specialized skills may be required to repair any breakdowns and integrate replacement parts into the system. CIMS can indeed be an iceberg of follow-on costs. The key to avoiding high follow-on costs and controlling maintenance costs lies in convincing the client to spend enough on the front-end planning and design of the CIMS.

In their haste to avoid follow-on costs, many adopters invest heavily in an inventory of spare parts and back-up systems. Amanda, Barbara, Tilly, Charlene, and Arlene all did so. Thus, Amanda and Charlene spent much more than they could afford to achieve their CIMS. In Amanda's case, the cost was offset by the high labor productivity. But since Charlene had no such advantage, their CIMS did not produce a good return on investment.

Fears about Reliability. This is a variation on the fears about maintainability. It is a fear common to the adoption of all-or-nothing, high-capital-investment technologies (Souder, 1987). Once a CIMS is installed, the entire plant depends on it. Any failure of the CIMS means downtime in production. Downtime translates directly into lost profits.

Any readers who have experienced the crash of their personal computer have a first-hand knowledge of this fear. When mysterious, undiagnosed stoppages occur and all attempts to alleviate them through conventional means fail, desperation soon prevails. The lack of diagnostic information is psychologically excruciating. How do you fix this problem so that you do not erase everything you have entered during the last hour? What is the source of the problem? Why are these strange characters appearing on the CRT screen? All the while, you fear that you may do something to worsen the problem, making recovery impossible. Imagine the feeling of the plant manager whose entire plant is caught in this type of bind! The possibility of this happening is prominent in the mind of every potential CIMS adopter.

To some extent, this fear is borne of unfamiliarity with the new technology. However, it is also borne of a real recognition that CIMS may require the addition of a very expensive cadre of experts who are smart enough to fix any failures on the spot, immediately. Unexplained stoppages do indeed occur with computers; this fear is real!

Low Tolerance for Critical Risks. All systems require some minimum level of investment, or *critical investment,* before they can achieve an acceptable payback. A proportionate level of risk accompanies this minimum investment. Let us call this the *critical risk*. The problem is to motivate the potential adopter to stay with the development long enough—that is, to assume the critical risk—so that benefits are derived. This was a problem at all four failure cases examined here.

This problem of the user's willingness to assume a critical risk is more complex than the above discussion may imply. The critical risk is not always apparent at the start of a project. Although the state of the art in CIMS engineering is highly advanced, it is still difficult to accurately predict the cost of a CIMS development. It is even more difficult to accurately estimate the minimum investment required in a CIMS development before any benefits can be achieved (the critical investment). Thus, the critical risk equation is fraught with considerable uncertainty. And uncertainty is precisely what the potential adopter wants to avoid.

There are two potential solutions to this problem: uncertainty absorption and better estimating. Uncertainty will be "absorbed" when the MSI accepts parts of the critical risk through warranties and cost-sharing arrangements. Better estimating and costing concepts will increase the potential adopter's confidence in the MSI, and hence in the CIMS.

The Too Simple/Too Complex Dilemma. Every potential user naturally questions the utility of anything simple. Simplicity carries an image of ineffectuality. This is especially true in an industry that is accustomed to noisy, large, massive, and mechanically complex equipment, where effectiveness is equated with dirt, noise, and size. If it is small, simple, and clean, it can't be much good. On the other hand, complexity carries an image of difficulty in use. If it is too complex, it can't be useful.

This dilemma characterizes CIMS, which is necessarily complex. Attempts to simplify it may in fact diminish their potency. The Peggy case affords an excellent example of this. Beta's attempts to simplify the CIMS to make it more acceptable and useful had just the opposite effect: The simplified CIMS was viewed as ineffectual and management canceled the work.

The client's felt need for back-up systems to ensure against failure often contradicts the client's desire to have simple, low-cost systems. Being attuned to client needs, the MSI is inclined to opt for the cheapest system in the short run, compromising long-run quality for short-run fail-safe features.

Islands of Technology. It is a common notion that the presence of any islands of old technologies (manual interfaces) will diminish the success of the CIMS, since paperless interfaces are the key to productivity enhancements with CIMS. This was the case for Peggy's CIMS and Arlene's CIMS, which were characterized by many islands. Personnel were required to physically read bar codes at several points within the process. Information was printed out by a computer and reentered by hand into another computer at several points within the processes. Human operators were required to watch various indicators and manually adjust conditions according to their observations. These types of "islands" raise the cost and diminish the benefits from CIMS.

Internally Contradictory User Goals. Job shops that want to fully automate are an inherent contradiction in terms and philosophies. Job shops constantly fiddle and jiggle to make their products work. The job shop culture is a cut and fit mentality, run by "seat of the pants" procedures. This makes them a difficult automation client. Yet every one of the eight cases studied here either had a job shop mentality, or was a pure job shop in structure and nature. The four successful cases were able and willing to move beyond their old job shop cultures.

Lack of Experience in Prime Contracting. When the client acts as its own prime contractor, there are no checks and balances. The client is often its own worst enemy in developing and implementing CIMS. An outside prime contractor usually has a more critical mind and a broader viewpoint. When the client is inexperienced in the role of prime contractor, as was the case at Stella, Peggy, and Charlene, coordinating the efforts of the subcontractors can become self-defeating.

A Nonparticipating Client. The literature strongly indicates that the potential adopter must be an equal partner in the CIMS development process (Cohen and Zysman, 1987). The client that makes no input during the project and just looks on from a distance is a major barrier to good CIMS design and development. The Tilly, Stella, and Peggy cases demonstrate the importance of user involvement. Tilly remained involved as an equal partner in the CIMS design and development work from the start. Stella and Peggy did not, and their CIMSs were much less effective. User involvement increases the chances that the CIMS will in fact meet the user's needs, that the user will be properly trained and acclimated to the resulting CIMS, and that the CIMS will in fact be what the user wants.

Desire for Turnkey Technologies. A turnkey CIMS is the great hope of the nonparticipating client. Three of the four failure cases examined here wanted turnkey CIMS. However, the state of the art in CIMS design and development is not yet at the point where an MSI can tailor a CIMS to a client's needs in the absence of periodic interaction and close involvement with that client. Moreover, the evidence suggests that labor forces seldom fully accept a turnkey CIMS where they have not been involved in the design and implementation.

Lack of In-House Manufacturing Engineers. In part, the need for turnkey manufacturing technologies is a result of the fact that most U.S. plants have laid off all their production engineers and plant software designers in an effort to cut costs. This makes them totally dependent on outside vendors. But, perhaps more importantly, the loss of in-house expertise hinders the MSIs because they have no manufacturing professionals with whom to interface in designing appropriate systems for the client. This was precisely what happened at Peggy and Arlene, where there were no in-house experts to help Beta and Omega specify the appropriate systems.

Client Over-Concern for Controlling CIMS Costs. If the client is obsessed with controlling costs and will therefore not permit any changes, this will severely constrain the work, which must naturally be flexible to meet changing events over the life of the project. The Peggy and Charlene cases illustrate this barrier.

Tenacious Owners. The personal virtue that accounts for the success of many small shops is the factor that now limits them: tenacity of the owners. Small shops built on grit and perseverance are enculturated with the conviction that these same characteristics will continue to conquer all.

Most service plants are extensions of the personalities of the founders. It is difficult for them to understand how a former asset can be the source of their current problems. The hypersecretive nature of small shop owners also helps to erect formidable barriers to the collection of the basic process information that the MSI needs to define the CIMS. These were exactly the problems that confronted Gamma at Charlene, where Charlene's managers were reluctant to share operating data and order flow process information with the MSI.

Unrealistic Expectations about CIMS. Most potential users view the benefits of automation in terms of the reduction in low-skilled labor. CIMS does indeed substitute for some low-skilled labor, but the effective use of CIMS also means higher costs for more highly trained labor. This lesson is apparent in the Amanda, Tilly, Stella, Peggy, and Charlene cases, where management initially had the impression that CIMS would dramatically lower total unit costs. But this did not happen; the major benefits from their CIMSs were intangible: increased quality, reduced re-work, faster time to market, and better control. The problem is that these benefits are very difficult to accurately predict in particular cases.

Paralysis by Fear. Many managers are well aware that the alternative to doing nothing is to close the plant. Still, they fear doing anything. This is a version of paralysis by fear: The fear of worsening the situation is more immediate than the

TABLE 9.4 Thirteen Ways the MSI Can Become More Effective

Salesmanship and project abilities
Role flexibility
Market analysis abilities
Dedicated MSI professionals
Timing of work flow
Piecemeal development capabilities
Ability to manage client expectations
Appealing to basic management needs
Solutions selling
Patience to live with the natives
Using shop-floor language
Using the domino adoption theory
Using counterpart implementation teams

fear of closing the plant. This phenomenon was apparent at Dolly, Stella, and Arlene.

Internal Status Differentials. Operations departments often suffer a low-status image within the firm, whereas engineering departments enjoy a much higher status. This gap sets up a severe barrier, since the engineering and operations departments generally take adversarial positions, with engineering advocating change and operations advocating the status quo. The engineering group usually sides with the MSI, giving the operations group the feeling they are being overwhelmed, thus hardening their already uncharitable attitude toward the MSI. This is exactly what happened at Amanda, Dolly, Stella, and Arlene, and to some extent at Tilly.

Internal Organization Climates. The Stella, Peggy, and Arlene cases demonstrate how a lack of trust between top and lower management can present obstacles to CIMS adoption. At Arlene, top management diffused their perceived risks of adoption by delegating the adoption decision to the lower ranks. This, in turn, caused the risk-averse lower ranks to withdraw and further resist adoption. Top management then asked for system redundancy and back-ups. The additional costs of these add-ons were charged to lower management's budgets. This caused lower management to become more resistant to adopting these more reliable and fail-safe systems!

The fear of adoption may be overcome by building in redundancies and safeguards, but doing so runs counter to the client's desire to minimize costs and maximize simplicity. Thus, adding safeguards may fuel the client's impression that adoption carries many uncertainties, considerable complexity, and many potential traps beyond the basic costs and risks. As the Arlene case demonstrates, the delegation of risk responsibilities is likely to be a one-way street. Risks may be delegated to the lower ranks, but credit is likely to be taken by the upper ranks. Knowing this, the lower ranks are not likely to try to be heroes or heroines. Thus, it becomes easier and safer for the lower ranks to simply refuse to adopt the CIMS.

9.5 RECOMMENDATIONS

The MSI faces a competitive situation in a complex, changing market. The customer base is very segmented, splintered, and individualistic. The customer is constantly becoming increasingly more knowledgeable and more able to critically evaluate alternative CIMSs. Manufacturers and vendors are forward integrating into the MSI business, new MSIs are entering the market every day, and many successful adopters are using their experience as the base for launching themselves into the MSI business.

How should MSIs conduct themselves in the face of this complex and dynamic environment? The cases revealed 13 ways in which MSIs may increase their effectiveness (see Table 9.4 and the following discussion).

Salesmanship and Project Abilities. A successful MSI must have experience-based abilities in contractor relations, project management, and implementation. A successful background as a project monitor and program manager is the best basis for becoming a good MSI. The Peggy case vividly illustrates this.

A successful MSI must also have experience-based abilities in customer definition of wants and needs. A background in direct customer selling is a good basis for learning this skill. The Tilly, Dolly, and Arlene cases demonstrate this. Contrary to conventional wisdom, these abilities may be more important than a deep knowledge of CIMS technologies.

Role Flexibility. MSIs must be able and willing to play a very aggressive consulting role. They must be able to condition the customer to take their advice. Customers who lead MSIs raise considerable danger that inadequate CIMS will result. In the case of an MSI-sponsored innovation, the customer is unlikely to know what is best; the Dolly case illustrates this. Omega was initially overwhelmed by Dolly's style, and progress was minimal until Omega recognized that it should take the lead. Because Omega feared the loss of the customer, it initially tended to go along with Dolly's suggestions. As it turned out, Dolly's suggestions were only advisory; Dolly had assumed the leadership role only because they felt the need to fill the leadership gap!

On the other hand, it is well established that users often originate many innovations (Souder, 1987, pp. 178–198, 216–237). Moreover, if the MSI is too forceful, the client may abandon the effort due to a perceived lack of participation opportunities. This happened at Stella and at Arlene. Thus, it seems clear that the MSI must be very perceptive about clients' personalities, must understand whether the client wants to lead or be led, and must act accordingly. Rules and methods for determining who should lead have been developed in Souder (1987).

Market Analysis Abilities. Successful MSIs invest the time and patience required to analyze the systems needs of those users who do not know what they want and will not know until they see some prototypes. Successful MSIs understand and gain appreciation for the fact that the resistances to adoption and the reserved attitudes that characterize potential adopters are necessary parts of the low-profit mentality of this business, where one small mistake can shut down an entire plant.

To develop and implement a true total system, the MSI must spend whatever time is required to help the client define its objectives before the work begins. Neglect of this step exposes the system to later failure, at considerably greater costs, inconvenience, and loss of image to the MSI. Events at Stella and Arlene testify to this.

Dedicated MSI Professionals. The most successful cases studied here (Tilly, Amanda, and Barbara) were characterized by a dedicated group of MSI software engineers who moved in and figuratively lived at the plant. They won the confidence of the casting superintendent and learned the details of that particular plant, so they could design the software system most appropriate for that particular plant and that culture.

Timing of Work Flow. The Stella and Arlene cases demonstrate that when the MSI starts the software development too late in the project, or exhibits uncertainties about the design of the system, the client may be inclined to terminate the more exotic aspects of the system. This can cause the product to be much less capable, which may in turn cause the client to want to terminate the entire effort.

Piecemeal Development Capabilities. Many clients either are job shops or exhibit the job shop mentality of focusing on each job as an independent entity. In this case, the MSI must be content to do a piecemeal CIMS, rather than sell the entire system all at once. For example, at Tilly, the cost accounting portion of the CIMS was purposely fabricated early in the CIMS development project. This was done to demonstrate the benefits of this piece of the system relative to Tilly's existing accounting system. The results were so dramatic that any remaining resistance to the CIMS immediately vanished.

Piecemeal development is not without hazards. Piecemeal CIMSs do not usually provide the proportionate benefits of the total systems CIMS; piecemeal CIMS may be negatively viewed by the client as inadequate solutions; and piecemeal CIMS may cost the client much more in the long run. The Peggy and Arlene cases illustrate this. It is the MSI's responsibility to fully explain the total versus piecemeal alternatives to the client, and to openly counsel the client on the costs and benefits of each alternative. Here is a case where the MSI must expend the time and effort required to patiently educate the client.

Ability to Manage Client Expectations. Perhaps the biggest problem facing an MSI is managing client expectations. Customers typically expect the dollar to buy more than it actually does. How does the MSI persuade the customer that its perceptions are wrong? How does the MSI convince the customer that the initially cheaper solution will cost more in the long run? How does the MSI convince the customer to make important decisions in a timely fashion so that costs do not escalate? These questions need much more research.

As the Stella, Charlene, and Peggy cases illustrate, customers may think they are more sophisticated than they actually are. They may feel they are competent to design a CIMS. The ever-increasing "do it yourself" articles in the trade journals encourage the notion that users can build their own CIMS. Some users may in fact be able to design their own CIMS, but there is the danger that they will be persuaded that CIMS design and implementation is much easier than it actually is. As one subject who was interviewed during this study lamented: "We suffer from the instant-brain-surgeon-in-one-easy-lesson syndrome."

MSIs may also have ill-founded expectations. They can sometimes have unrealistic expectations of what they can accomplish in a given time frame and exaggerated expectations of their own capabilities. They may underestimate the difficulties of coordinating their subcontractors and the technical difficulties of the job. Alpha, Gamma, and Omega all committed errors of expectation in the Stella, Charlene, and Arlene cases, respectively.

Managing client expectations is a complex matter than can be resolved only through the creation of more intelligent and better educated users. MSIs should understand that one of their important roles is educating the world. This carries a social responsibility that goes beyond their immediate commercial goals. In the short run, MSIs may have to sacrifice some quick profits and devote conscientious time to counseling and educating their potential clients.

Appealing to Basic Management Needs. All the cases studied here illustrate how management's need for control and performance feedback information are the basic motives for adopting CIMS. Appeals to these motives are the key to adoption. All managers have a basic need for control information and for consistency of operations.

Solutions Selling. All the CIMS technology is available and applications know-how is becoming more commonplace. Knowledge and understanding of the customer and the ability to provide the solution to the client's problem are keys to CIMS adoption. The biggest barrier is knowing the customer's needs and requirements, so that the system can be designed to consistently and economically meet these needs. Gamma's efforts at Tilly are an excellent example of solutions selling. In contrast, their efforts at Charlene are a good example of poor solutions selling.

Patience to Live with the Natives. As a prerequisite to adoption success, the MSI must carry out an anthropological mission by living with the natives, at their plant, in their environment. This achieves two purposes: acquainting the vendor personnel with the technical and cultural aspects of the plant and winning the confidence of the natives.

As the Barbara, Tilly, Peggy, and Charlene cases show, success in CIMS implementations varies with the regard the client and the vendor have for each other. The formation of a true marriage partnership between the vendor and the client can be a very important factor in implementation and adoption success. The

MSI's reputation can trade off for higher costs and a more powerful CIMS in the adoption decision.

Using Shop-Floor Language. CIMS must be explained in terms that the key shop-floor personnel can understand. The shop-floor level does not control the adoption decision; however, as the Tilly and Charlene cases show, shop-floor personnel are the key to long- term successful implementation. The participation of shop-floor level personnel in the design and implementation of the CIMS promotes its long-term adoption and successful use.

Using the Domino Adoption Theory. Typically, an adoption situation is characterized by early adopters and followers. The interviews indicated that this is indeed the situation within the service manufacturing sector. The Tilly, Stella, Peggy and Arlene case interviews suggested the presence of some earlier adopters that they emulated. In addition, several independent consultants, systems suppliers, and trade groups were cited who played the role of circuit riders. These entities move freely throughout the industry, giving advice and promotive talks about CIMS, and generally proselytizing for the adoption of CIMS. Although the evidence is all anecdotal, there were strong suggestions that these circuit riders have a positive impact on adoptions. Thus, if the early adopters can be located, convincing them to adopt is the key to industry-wide adoption.

Using Counterpart Implementation Teams. The Barbara and Tilly experiences indicate that a key mechanism for CIMS adoption is matched counterpart personnel teams from the client and the MSI. Every function (engineering, production, training, cost, computers, and process) should be represented by at least one person on each team and all personnel should be full-time team members. The team members should continuously interact, and joint team meetings should be held weekly. The teams should report directly to their respective managements.

The Barbara and Tilly cases suggest that these teams may be more effective if they do not have a history of working together. Since every CIMS implementation is unique and creativity is desired, the use of previously formed teams may bias actions and thoughts. However, this idea is counterintuitive. Much more study is needed on this aspect.

9.6 CONCLUSIONS

Fenced in by outmoded equipment, old cultures, and worn-out facilities that once served them well, America's service manufacturers have become trapped in their own infrastructures. The tenacity of the owners, their handicraft management styles, and the hands-on production skills that once made them a potent force are part of the same dialectic that now is destroying them.

As America's service manufacturers go, so go America's military, consumer durable, and industrial repair capabilities, which may have come to depend on this sector. Federal assistance may provide some economic relief, but it is clear from this study that much more than economic relief is needed. The role of the MSI is critical to overcoming the organizational, social, cultural, and managerial barriers that effectively inhibit the adoption and diffusion of the CIMS, which can resurrect this sector. The care and feeding of MSIs is thus essential to America's repair and maintenance capabilities. Thus, it is important to study MSIs, to learn about their modus operandi, and to find ways to promote their effectiveness.

The results from the case analyses revealed seven factors that promote the successful adoption of CIMS. When the user has champions for the CIMS among both its top management and shop-floor ranks, and when the user has a well-developed strategic business plan, the successful adoption of CIMS is enhanced. Family ownership of the business and long-term employment stability were also found to enhance adoption. When the labor force for the CIMS was rigorously selected according to empirically defined performance and human personality standards, the CIMS was highly successful. When the labor force served internships at vendor facilities, the CIMS was much more successful. Surprisingly, the economic strength of the user was not found to be relevant to the success or failure of the CIMS.

Thirteen actions were found that the MSI can take to foster the successful adoption and diffusion of CIMS. Salesmanship, role flexibility, improved market and user analysis research, and the dedication of MSI professionals to user problems were noted as important actions. The willingness to promote the piecemeal adoption of CIMS, the patience to live with the customers and study their needs within their environments, and the willingness to learn the user's shop floor language were also found to be important factors. Gaining rapport and respect, managing client expectations, and skill in solutions selling also distinguished the successful MSIs. The most successful MSIs used a domino strategy of selling to the industry leaders first, which then made the sale to the followers much easier. Perhaps the most important attribute of the successful MSI was competence in project management. The most effective MSIs in this study were able to schedule and coordinate many subcontractors, and make very productive use of counterpart MSI and client teams.

This study has resulted in four major conclusions of interest to managers of service manufacturing firms and to managers of MSI efforts. First, service sector manufacturing personnel are not inherently resistant to the adoption of CIMS. Rather, they fear the loss of security and the confrontation with the unknown. To overcome this barrier, the MSI should try to make the unfamiliar more familiar. This can be achieved by implementing the CIMS in piecemeal fashion and by involving the user personnel in setting the standards by which they will be judged. In this approach, the personnel try out each portion of the CIMS, going on to the next only after the previous portions are operating smoothly. The relevant personnel fully participate in setting the output standards and productivity targets for each portion. Note that this participative approach is not the same as involving the personnel in the selection and design of the CIMS, which is also a potent technique for overcoming resistance. Rather, in the approach recommended here, the output standards and productivity targets determine how the performance of the personnel will be judged and how they will be financially rewarded under the new technology.

Second, continuous training is essential after the CIMS comes on stream. Technical training is essential to familiarize the personnel with the new innovation. Moreover, training plays an essential social role. The peer-to-peer and novice-to-expert social interaction that occurs in most training sessions is a valuable mechanism for opinion exchange, opinion formation, and team formation.

Third, constant experimentation with the CIMS must be encouraged. Personnel rotations, multiple-skill development for personnel, meetings where new ideas are discussed, and suggestion programs are all very important. The CIMS must be an open system that grows and flexes and is amenable to add-ons. The important thing is to let the operators know that the system is open ended and that suggestions can be implemented.

Fourth, the operators and plant personnel must be encouraged to talk about the CIMS as a common part of their daily lives. Encourage them to talk to visitors, play the role of tour guides, and generally treat the CIMS like an everyday item.

As a new and evolving institution, the MSI movement needs help. No institution can survive for long without an infrastructure. MSIs need to form a trade or professional association (e.g., an international institute of systems integrators), to lend more credibility to their work and to attract the necessary human resources. Young high school graduates need to be encouraged to enter the manufacturing sciences and to consider careers in systems integration. There is a strong resurgence of interest in manufacturing among the university community today, which MSIs could link with via joint ventures, direct research grants, and other collaborations.

9.7 CHALLENGES FOR FURTHER WORK

This study has been only a preliminary examination of a rapidly evolving new institution that may have dramatic effects on America's manufacturing capabilities. Much more research is needed on the effectiveness of the MSI and the means for overcoming the barriers to the adoption of new technologies among America's service manufacturers.

This study has resulted in many preliminary findings for further testing and integration. The following is an inventory of hypotheses and propositions developed from this study that can be used to guide further studies of this nature.

Hypotheses and Propositions for Further Work

1. Inside engineering groups are prime movers for the adoption of CIMS, but only if they occupy a high status level within the parent firm.
2. Whether the CIMS is state of the art is not so large a concern as whether the system "fits" the culture and the ability of the personnel to use it. *Corollary:* Attempts to advance the plant with radical new systems may fail; the system should be no more advanced than the plant.

3. For success in the adoption of CIMS, champions are required at two organizational levels of the adopting organization. There must be at least one high-level champion and one shop-floor champion within the adopting organization.
4. When the economic situation of the plant becomes so bad that the adoption of CIMS becomes imperative for survival, the fear of failure and the associated regrets may also become so great that they effectively prevent adoption when the plant needs it most.
5. The fear of adoption increases as the plant becomes increasingly older, less competitive, and financially disadvantaged. *Corollary:* Target plants for renovation and upgrade when they are still profitable, not when they need it most.
6. The fear of adopting new things is least when profits are high and greatest when profits are low.
7. Although the fear of adoption may be overcome by building redundancies and safeguards into the CIMS design, this runs counter to the client's desire to minimize costs and maximize simplicity, thus adding to the client's impression that adoption carries many uncertainties due to complexity and many cost traps beyond the basic costs and risks.
8. If top management asks for system redundancy and back-ups, lower management may find the costs of these add-ons charged to their budget, so that lower management may actually be more resistant to adopting more reliable and fail-safe systems.
9. Top management may diffuse their perceived risks of adoption by delegating the adoption decision to the lower ranks. This, in turn, may cause the lower ranks to withdraw and further resist adoption.
10. The delegation of risk responsibilities is likely to be a one-way street; risks may be delegated to the lower ranks, but credit for the success of the implementation is likely to be taken by the upper ranks.
11. There are key low-level decision makers in any organization whose support, once won, can become an important promotive factor in adoption.
12. CIMS must be explained in terms that the key shop-floor personnel can understand.
13. The vendor must "live with the natives"; this achieves two purposes: acquainting the vendor personnel with the technical and cultural aspects of the plant and winning the confidence of the plant personnel.
14. The degree of success of CIMS implementations varies with the amount of front-end planning; a CIMS that has been well planned is more likely to succeed.
15. The probability of success of CIMS implementations rises with the degree of regard the client and the vendor have for each other.
16. The degree of success of CIMS implementations varies as the degree of regard and the amount of technical planning increase together. When either

the degree of regard or the amount of technical planning are disproportionately low, the implementation will suffer.

17. The systems integrator's heavy emphasis on front-end planning will naturally be resisted by most users since users are proactively oriented to early completion of the implementation. Thus, most MSIs will need to sell the prospective user on the need for heavy investments in planning.

18. The use of CIMS makes it much more profitable to manufacture within the United States to fewer CIMS users manufacture offshore.

19. The formation of a true marriage partnership between the vendor and the client is a key factor in implementation and adoption success.

20. Technology is not a barrier to CIMS success; we have all the technology we will need. Rather, solutions selling is one of the barriers to CIMS success, so the most successful MSIs must have a good solutions selling staff.

21. The major benefits from CIMS are intangible: increased quality, reduced re-work, faster time to market, and better control. These are more difficult to estimate and understand than cost reduction.

22. Because firms cannot or are not willing to accept the risk of downtime, they will only buy a CIMS that they are sure will work, from whomever they feel will provide as near to 100% perfect service and reliability as possible.

23. Adoption depends on the degree to which the firms have managers who are involved in extraorganizational professional activities, since these activities provide many opportunities to hear the pros and cons of new equipment debated and to trade experiences with peers. This results in a diversity of viewpoints, theses and antitheses that can be synthesized into viewpoints and positions, and a more conceptual perspective of the firm's total environment.

24. New facilities are easier to automate, because they don't have old cultures that resist new things. Older facilities have supervisors that are trained in older cultures.

25. New facilities are also easier to automate because management always wants the best for their new plants.

26. Shop-floor participation in the design and implementation of the CIMS promotes its long-term adoption and successful use.

27. One barrier to CIMS adoption is not the technology, but knowing the customer's needs. The technology is well defined and available to the world; it does not require much sophistication or technical know-how to assemble a CIMS. But the MSI *must* know the customer's needs and requirements so that the system can be designed to consistently and economically meet these needs.

28. The second biggest barrier is managing expectations. MSIs always have expectations that exceed reality in terms of what can be done in a given time and for a given dollar.

29. The third biggest barrier is managing the client. Most customers think they are more sophisticated than they actually are. They think they know what they need and they think they know how to design a CIMS. This is due to the many articles that say how easy it is to design a CIMS.
30. Another barrier is managing the client's risk aversion. This is closely related to proposition 28. How do you convince the customer to make important decisions and not permit things to hang for a long time, perhaps long enough that costs increase or the system becomes obsolete?
31. The client always fears that the MSI will complete the development and then disappear with no conscience about helping the client follow-on or maintain the system.
32. The combination of family ownership and a heritage of a quality specialty product is a potent force for adoption. Family ownership gives a firm the latitude to look far into the future, since the firm is not compelled to answer to an annual profit cycle or quarterly dividends.
33. When the entire management team consists of sons, daughters, grandsons, and sons-in-law of the firm's founder, the firm is used to doing things by consensus, the family way. Therefore, the decision to adopt will be time-consuming, but once that decision is made, it will be very permanent.
34. If the client has well-defined wants, well-defined work goals, and a well-defined time horizon, the CIMS design and adoption will proceed much more smoothly.
35. It is important to implement CIMS in piecemeal fashion (e.g., develop and try out the cost accounting portion of the CIMS) early on, so that the customer can gain some idea of how beneficial the entire system will be.
36. It is better for the client to grow its development and implementation taskforces internally than to assemble a previously experienced team, since each CIMS is different and prior taskforces are likely to exhibit many biases.
37. A client's lending of its own personnel to vendors is an effective way to aid in the transfer and training for CIMS.
38. The client's corporate plan and strategic market analysis are key success factors. If they are well developed, the client and the MSI can assemble a joint team and conduct a complete needs analysis, a thorough technology survey, and a systematic equipment selection, then make accurate budget projections.
39. One key mechanism for CIMS adoption is matched counterpart personnel teams from the client and the MSI. All the necessary functions (engineering, production, training, cost, computers, and process) should be represented. They should be allowed to grope their way through the details and be a naturally emerging team. That way, there are no prior biases. The task forces should continuously interface with all the represented functions in an interlocking fashion. The teams should report directly to higher management.

40. A fully integrated CIMS will be the most successful, but will also carry the greatest risks of failure.
41. It is vital to involve all the affected parties at the start if the objective is to develop and implement a true total system.
42. Implementation requires an internal marketing campaign. The adopters are more likely to adopt when their supervisors, their peers, and their own "clients" within the organization support their adoption and use of the innovation.

SUMMARY

This chapter has presented the requirements for adopting computerized integrated manufacturing systems. The role of manufacturing systems integrator (MSI) is discussed as well as specific case studies on the successes and future of CIMS. The lesson learned provides useful guidelines to management, systems designers, and users in successfully managing the implementation of CIMS.

ACKNOWLEDGMENTS

The author expresses sincere gratitude to Dr. William. E. Souder, Professor of Engineering Management and Industrial Engineering and Director of the Technology Management Studies Institute at the University of Pittsburgh, for providing the case studies presented in this chapter. The studies were funded by grants to the Technology Management Studies Institute from the Center for Information Technology at Northwestern University, from the Alcoa Foundation, and from the Pennzoil Company.

QUESTIONS

9-1 Discuss the role of a systems integrator in a high-technology manufacturing environment.

9-2 Outline the risks involved in implementing CIMS.

9-3 What are the advantages and disadvantages of implementing a turnkey CIMS?

9-4 How should users of CIMS evaluate the total benefits and costs of the system?

9-5 Discuss six conditions that could retard the adoption of CIMS.

9-6 What is the role of a systems integrator in front end planning work required for CIMS?

10
TECHNOLOGY MANAGEMENT CASE STUDY: GENERAL ELECTRIC HIGH-SPEED HORIZONTAL PROJECT

> This chapter is a case study contributed by Professor Paul S. Adler in collaboration with Professor Steven C. Wheelwright as the basis for class discussion rather than to illustrate either effective or ineffective handling of an administrative situation. General Electric Company proprietary data have been disguised. The material was included with permission.

In March 1985, Matt Steele had reason to feel good about General Electric's first high-speed horizontal (HSH) project. The new HSH line was producing fluorescent lamps in GE's Bucyrus, Ohio, plant at a rate of 6000 units per hour versus a maximum of 3500 units per hour produced by the old horizontal line. At a cost of over $15 million, the HSH project had been one of the major successes of 1985 for GE's Lighting Business Group (GE-LBG). A second, similar HSH line that would bring speeds to an unprecedented 7000 units/hour had just been approved for Bucyrus.

As general manager of manufacturing for the LBG's fluorescent and high-intensity products, Steele saw these two new production lines as assuring efficiency in the high-volume, standard 4-foot, 40-watt (F40) fluorescent lamp operations. Additional HSH lines planned for the Circleville, Ohio, plant would need to be modified to support a long-range manufacturing strategy of focusing that plant on the lower volume, specialized fluorescent lamps of varying diameters and lengths. But before work began on the appropriation request for Circleville, Steele decided to review that "focus" policy and the technology strategy associated with it.

10.1 BACKGROUND

More than 400 million fluorescent lamps were sold in the United States in 1984. (Figure 10.1 explains the components of fluorescent lighting.) With only three

FLOURESCENT LAMP PARTS

Basically, a flourescent lamp is made up of the following components: 1. a glass tube, or bulb, internally coated with flourescent material called phosphors. 2. Electrodes supported by a glass mount structure, and sealed at the ends of the tube. 3. A filling gas to aid starting and operation—usually argon, or argon with neon. 4. A small amount of mercury which vaporizes during lamp operation. 5. A base cemented on each end of the tube to connect the lamp to the lighting circuit.

In order to produce fluorescent lamps of high light output, long life, and good maintenance of light output through life, careful control must be maintained over the selection, purity and assembly of all the components of the lamps. Some of the prinicipal characteristics and requirements on the components are these:

Tube. Acts as airtight enclosure for the mercury, the filling gas, the cathodes, and the phosphor coating. Glass must be free of structural defects and cleaned before lamp assembly.

Bases. Connect the lamp to the electric circuit, and support the lamp. Lamps for preheat and rapid start circuits use two contacts on each end of the lamp. The bipin base is used on all preheat and many rapid start lamps. Some rapid start type lamps, such as high output and Power Groove® lamps, use recessed double contact bases because of the higher ballast voltage required with lamps longer than 4 feet. Instant start lamps require only one electrical contact on each end of the lamp; thus the single-pin base is most commonly used. Some instant start lamps use bipin bases with the two contacts connected together inside the lamps.

Mount Structures. Close off ends of the tube and support each cathode. Wires leading from base are sealed off here. These wires are made of special metal, called Dumet wire, which has virtually the same coefficient of expansion as glass. The mount structure also includes a long glass exhaust tube; during manufacture, air is pumped out of the bulb, and the filling gas and mercury are inserted. The exhaust tube is then cut and sealed off so that it fits inside the base.

Cathodes. Cathodes provide terminals for the arc and a source of electrons for lamp current. In some lamps they function alternately as cathodes and anodes, but are commonly called cathodes. In other lamp designs, separate anodes are used because they best fit lamp design requirements. Plate anodes in High Output lamps and wire anodes in Power Groove® lamps are used to reduce the wattage loss at the lamp ends. Cathodes are usually made of coil-coiled, triple-coiled or stick-coiled tungsten, like an ordinary lamp filament, except coils are filled with alkaline-earth oxides. These oxides emit electrons more freely, thus minimizing losses and keeping efficiency high.

Mercury Vapor. Droplets of liquid mercury are placed in the fluorescent tube during manufacture. During lamp operation, the mercury vaporizes to a very low pressure (about 1/100,000th of atmospheric pressure). At this pressure, the current through the vapor causes the vapor to radiate energy most strongly at one specific wavelength in the ultraviolet region (253.7 nanometers). Higher mercury pressures tend to reduce the production of this ultraviolet line. The mercury pressure during operation is regulated by the temperature of the bulb wall.

Filling Gas. Besides mercury, the tube also contains a small quantity of a highly purified rare gas. Argon and argon-neon are most common, but sometimes krypton is used. The filling gases ionize readily when sufficient voltage is applied across the lamp. The ionized filling gas quickly decreases in resistance allowing current to flow and the mercury to vaporize.

Phosphor Coating. Transforms 253.7-nanometer radiation into visible light. The fluorescent lamp gets its name from the fact that the phosphor coating fluoresces. The chemical make-up of the phosphor determines the color of the light produced. Phosphor particles in fluorescent coatings are extremely small—approximately 0.0007 inch in diameter. Careful control of phosphor particle size in necessary to obtain high lamp efficiency.

Auxiliary Equipment

Ballasts

The principal function of a ballast is to limit current to a fluorescent lamp. A ballast also supplies sufficient voltage to start and operate the lamp. In the case of rapid start circuits, a ballast supplies voltage to heat the lamp cathodes continuously.

A fluorescent lamp is an arc discharge device. The more current in the arc, the lower the resistance becomes. Without a ballast to limit current, the lamp would draw so much current that it would destroy itself.

An inductive ballast constitutes the most practical solution to limiting lamp current. The simplest inductive ballast is a coil, inserted into the circuit to limit current. This works satisfactorily for lower wattage lamps. However, for most flourescent lamps in use today, line voltage must be increased to develop sufficient starting voltage. Also, rapid start circuits require low voltage to heat the electrodes continuously. To meet these requirements, most ballasts today contain transformers.

FIGURE 10.1 Explanation of the components of fluorescent lighting.

other serious competitors—GTE (Sylvania), North American Philips (which had taken over Westinghouse's U.S. lighting operations in 1983), and Duro-Test—GE had supplied over one-third of the fluorescent lamp market in recent decades.

End users of fluorescent lamps were supplied by electrical distributors (who sold 63% of the units in 1984), by original fixture equipment manufacturers (14%), hardware stores (7%), and discount, variety, and department stores (3.5%)[*]. With a base of over 1 billion installed lamps in the United States, it was estimated that over three-quarters of annual fluorescent lamp sales were replacement units; less than one-quarter of sales were destined for new sockets. Lamp manufacturers were sheltered from the business fluctuations typical of many other major industries because of the stability and size of this replacement market.

Technical changes in fluorescent manufacturing traditionally had been slow but steady. Process speeds gradually had evolved from a level of approximately 1500 lamps per hour in the 1950s to about 3500 in 1980. Although not all firms in the industry were equally strong in manufacturing, equipment technology was fairly evenly spread. And given the importance of the replacement market, there was considerable incentive in the industry to keep new products compatible with existing sockets. Westinghouse and Sylvania had seemed content to follow GE's lead in this during the 1970s.

The moderate pace of product and process innovation had made distribution a key determinant of GE's market share. GE sold its products primarily through distributor houses, which in turn sold to electrical contractors and consumers. According to the 1984 Census there were 5000 to 7000 distributors in the United States. The National Association of Electrical Distributors, based in Stamford, Connecticut, had 800 member companies representing some 3000 branches, and the association had a list of some 400 prospective members.

GE's market position was enviable: its products commanded a premium price from brand-loyal customers, making GE distributorships a profitable proposition form both parties, and there was something of a trade-off between GE's strength in the distributorships and its relatively lower penetration of the original equipment manufacturer (OEM) distribution channel. Price, not lamp quality, was critical to the OEM market. However, that market represented not only substantial sales volume but also a potentially valuable source of new product ideas. OEMs were important too as allies in creating new sockets for new lamp products.

GE-LBG had adapted to its particular environment by expanding the number of its lamp departments from 3 to 12 between 1954 and 1982, to better serve major segments of the market. Each was responsible for its own process and product development, and each was a profit center.

10.2 THE WINDS OF CHANGE

Signs that things might be changing in the lighting industry emerged in the wake of the 1973 oil price rise. GE's new "Watt Miser"® products, lamps that reduced

---
[*]Data supplied by the National Electrical Manufacturers Association, Washington, D.C.

energy consumption per lumen delivered, were brought to market in the mid-1970s. But slower growth in this period meant that controlling costs was more important than product innovation. Processes were tweaked, head counts in the plants reduced, and numerous improvements made. While the consumer price index more than doubled, the price of a standard F40 lamp increased by about 50% (in current dollars) between 1970 and 1980. The energy crisis had served to reinforce a focus on costs that made product innovation appear problematic.

In 1980, Osram, a key player in the European lamp market, had considered acquiring Westinghouse, one of GE's principal competitors. In 1983, however, North American Philips completed the acquisition of Westinghouse's lamp business, boosting its U.S. market share from between 5% and 6% to some 20% and giving it a solid base in the OEM market as well as a foothold with distributors. As Europe's leading lighting products supplier, Philips (Europe) had long pursued an aggressive technology policy. Although its lighting business was somewhat smaller than GE's, Philips' Eindhoven research facility was "perhaps the most elegant and the most advanced in the world" when it came to lighting research. Through its work there, Philips had developed a considerable reputation for new product initiatives and manufacturing efficiency.

Given the high-volume nature of lamp manufacturing, the production speed was a good indicator of how effective a production process a manufacturer had and to

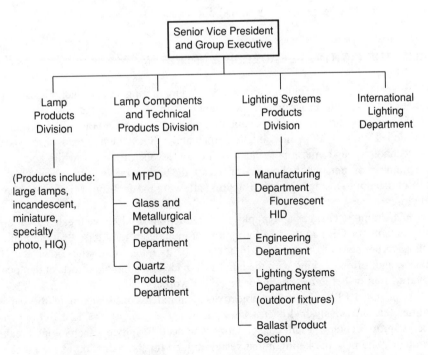

FIGURE 10.2 General Electric Lighting Business Group operating chart.

Evolution of product/market drivers for fluorescents

	Past → Today	Transition (0-3 years)	Future (3-5 years)
Market drivers	Initial cost	Cost of light	Total building/system efficiency/flexibility
Fixture	Housing only	Better quality light	Integrated modular fixture/lamp/ballast/ controls/ceiling Design approaches
Lamp Ballast	Standard products	Watt Miser products electronic ballast	
Controls	Dimming (at best)	Programmable controllers	
Ceiling Design	Suspended grid	Modular	

FIGURE 10.3 Evolution of product market drivers for fluorescents.

what degree it was controlled. In 1981, when GE-LBG had analyzed the lamp industry's production processes and the estimated speed of equipment for each of its major worldwide competitors, it had not ranked itself very near the top.

10.3 THE LIGHTING LEADERSHIP PROGRAM

GE's response to these findings focused both on marketing and production costs. Its Lighting Leadership Program (LLP), announced in 1983, targeted cost reductions of 15% to 25% through plant consolidation and process automation. The program envisioned closing 10 plants and retaining about 30, which would be focused by technology and mission to achieve higher levels of output while reducing the number of personnel. For example, all the industrial F40 lamps would be consolidated into a single plant at Bucyrus, allowing the Jackson, Mississippi plant to close.

Such dramatic changes and employment reductions could have created a variety of problems for GE had the LLP program not included a long-term "people plan." GE gave notice up to three years in advance of its intention to shut down the 10 plants and, often working with the unions, it made considerable efforts at the local level to find or help create jobs for layed-off workers.

Achieving LLP's aggressive objectives required a consolidation of the capabilities that previously had been somewhat dispersed among the 12 LBG departments (profit centers). Plant consolidation, the need for process technology breakthroughs, and new marketing initiatives all argued for the reduction of the number of product departments and a more functional organization. By 1985, the main

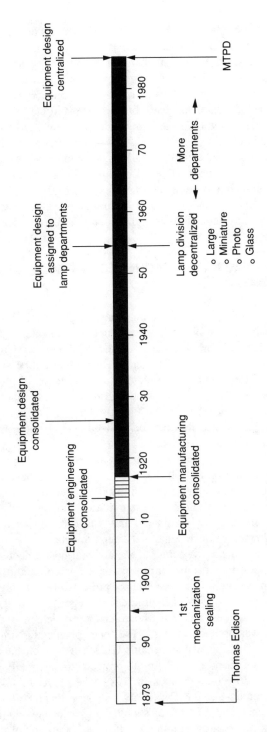

FIGURE 10.4 Evolution of equipment manufacturing at LBG.

lamp product responsibilities were grouped into commercial and industrial (LSPD) and consumer and retail (LPD). (See Figure 10.2 for 1985 LBG operating chart.)

Ron Mathewson, LSPD's general manager of marketing and sales, saw his challenge and opportunity that year as going beyond increased sales and cost reduction. As energy-cost escalation slowed, he saw nonenergy criteria becoming progressively more expensive. The early breakthroughs in the Watt Miser line had "mined" much of the easily available lamp-efficiency gains. On the positive side, new lamp phosphors offering better color, new lamp sizes and shapes, and more sophisticated energy-saving, electronic ballasts seemed to be opening up array of new, high-value product opportunities. Lighting controls could be from simple dimmers to programmable systems, and fixtures, the EOMs' domain, could help provide better "quality." A new world of fixture/lamp system designs for greater efficiency and flexibility seemed to be right around the corner (see Figure 10.3).

10.4 THE MANUFACTURING TECHNOLOGY PROGRAMS DEPARTMENT (MTPD)

MTPD had been formed in 1981 to centralize the design and building of equipment for LBG, previously the responsibility of product departments. Through this new department, LBG hoped to develop the focus and depth necessary for the dramatic process improvements that LLP had targeted. Harnessing the potential offered by electronic manufacturing technologies required the formation of a critical mass of electrical engineers and other experts whose efforts could be applied to suitable projects. MTPD drew some 50% to 60% of the lamp department home offices' manufacturing engineers into the new centralized organization still leaving the plants' manufacturing capabilities intact.

Between 1910 and 1920, lamp equipment engineering had first been consolidated; it remained centralized until 1954 (see Figure 10.4). Seeking the benefits of product-focused missions, the "lamp division" at that time was decentralized into four departments, with equipment design divided among them. In the ensuing years, division decentralization continued, and with it the dispersion of manufacturing engineering resources over time to 12 departments.

Encouraged by a strategic objective of incremental cost reduction, lacking the critical mass to support in-house major process development, and in the absence of any market challenge from a rival like Philips, projects of relatively limited scope were the rule. As one of the MTPD program managers put it, it was a culture where "you gave an engineer a drawing board and a few helpers and he'd call you when the machine was up and running." The focus of manufacturing engineering was incremental cost cutting, which became GE's hallmark of that period.

But there had not been enough small projects to keep GE abreast of its competitors. This had become apparent in 1979 when LBG found it had to turn to Hitachi for equipment for the Circleline™, or ring-shaped, fluorescent lamp. LBG's first experience with an outside supplier for major new equipment proved to be very successful and allowed a quick, efficient "ramp-up" of production.

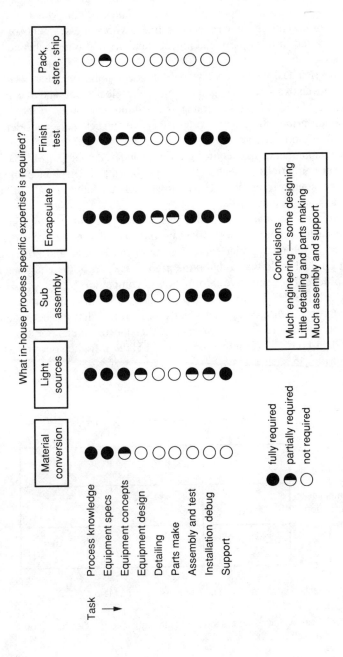

FIGURE 10.5 Assessment of MTPD Process Skills—1981.

With the formation of MTPD, such make/buy equipment decisions could be approached more systematically. Rather than buy entire production lines, MTPD's general manager wanted to use outside vendors to leverage MTPD's capabilities and give his department an opportunity to learn. Thus, MTPD was to serve as a prime contractor on entirely new production lines, with some segments of the line being subcontracted to third-party equipment suppliers and other subcontracted to MTPD groups.

One of the first tasks was to draw up a process skills inventory (see Figure 10.5). Detailed analysis showed gaps in both the electronics and the materials handling areas. MTPD adopted a strategy of building capability in the former as fast as possible; in the latter, where proprietary advantages were considered rare, MTPD would rely on vendors.

For the development of new equipment for incandescent lamps, MTPD decided to pursue a very ambitious project (called PRO-80) with little reliance on external suppliers. The equally ambitious HSH project, on the other hand, brought an outside vendor into a central role.

The key to the new make/buy policy was to recognize that in the fluorescent segment the basic technology was 40 years old. There was no great danger, it was believed, in letting suppliers do the casting, machining, and other standard steps. What GE had to do was to build the proprietary part of the equipment while avoiding "spillovers" of sensitive information to competitors.

Speaking of the HSH project, Dick Hunder, the HSH program manager explained why it was not conducted strictly in-house, "That project would have taken us hundreds of man-years—we had neither the resources nor the calendar time to do it in-house." (Figure 10.6 shows MTPD's framework for assessing program priorities in such cases.)

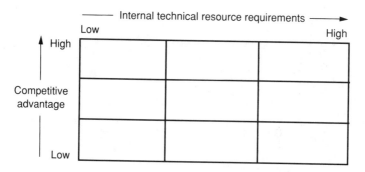

- Competitive advantage = $f(\Delta \text{Lead time})^{-1} \times (\Delta \text{Margin})$

- Technical resources = person years of thought leaders (electronic and mechanical) plus expense dollars.

FIGURE 10.6 Selection of program priorities versus application of limited technical resources.

The centralization of equipment designing and building skills in MTPD had created some tensions. Plant managers were, more than ever, being evaluated on their cost reductions but no longer had the manufacturing engineering resources needed to do all the projects that could generate these improvements, and MTPD could not always accommodate their needs for help. Therefore, a Users Council was formed of the manufacturing managers served by MTPD. Its mission was to raise and address issues of organizational interfaces, skill mix requirements, project management procedures, and MTPD equipment quality. MTPD welcomed this initiative, since experienced plant people played an important role in the design review teams.

10.5 THE HIGH-SPEED HORIZONTAL PROJECT

The HSH project, the core of LSPD's response to new competitive pressures and senior management's expectations under the Lighting Leadership Program, was a pathbreaking project in several respects. Technically, it would provide GE with the lowest costs in the fluorescent lamp industry by almost doubling the speed of the manufacturing line. Over some 20 years, GE had almost exhausted the improvement opportunities of indexing machines (in which each item in each step of the production line was stopped before being worked on) by pushing the speed from 1500 to 3500 units per hour (see Figure 10.7). HSH, operating instead in a continuous process (each item was worked on while in motion) would allow 6000 units or more per hour—a dramatic change of pace for GE's engineers and manufacturing people. (See Figure 10.8 for the basic steps in the fluorescent lamp

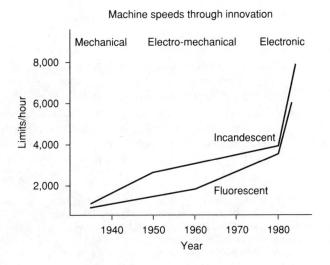

FIGURE 10.7 LBG productivity and automation (Source: General Electric—MTPD).

manufacturing process.) With such speeds, two HSH lines could replace 14 vertical indexing lines at the Bucyrus and Jackson plants, and allow cost reductions of over 15%.

HSH also represented only the second time that GE had purchased the key components of a new production system from an outside source, in this case from CKD, a Japanese company (see Figure 10.9 for data on CKD), and thus the project was highly visible. And unlike the Hitachi equipment it was not off-the-shelf equipment that was purchased for HSH. This decision was not made hastily; central to it was the fact that CKD had recently installed an earlier generation system at one of Europe's major firms. After a series of problems (due primarily to the quality of component materials), CKD's unit had passed its acceptance test running at 4500 units/hour. No other supplier could promise as much.

The knowledge embedded in CKD's equipment was basically new to LBG. MTPD and LSPD took responsibility for some of the HSH process definition steps, but CKD was responsible for some 60% to 70% of the system. CKD would provide the central "group" on the HSH, leaving coating, mounts, and basefill to GE (see Figure 10.8). (The distinction between the line and its central group dated from the vertical process where the nongroup components were not integrated into the line.) MTPD input of ideas and components would help bring the equipment capability from the 4500 units/hour initially offered by CKD to 6000.

The sensitive, proprietary areas of GE's contribution were the vacuum technology and the quality-control system. GE could easily protect the former, where it had a real advantage, by contract language. The quality system's value, on the other hand, was not so much in the data processing as in GE's ability to interpret its output.

Reflecting MTPD's new make/buy policy, an automatic packaging system at the end of the HSH line would also be subcontracted. None of GE's competitors had installed such systems. The task was complex, since packaging materials did not have the rigidity that enabled easy, automated handling. GE initially turned to Pyramid, a company based in Vancouver, B.C., for this work; but in 1983 that company went bankrupt, and CKD was asked by GE to complete that system.

Working with CKD had been an instructive experience for GE. An entire year of discussions took place before the Japanese company felt ready to sign a contract[*]. While their people were not legalistic in their approach, they were much more concerned than GE with detailed knowledge of processes, materials, and performance expectations before committing themselves to the project. Their questions had the effect of forcing GE to better define the manufacturing process and its specifications up front. Indeed, when it came time to sign the contract, GE

---

[*]The contract negotiations revealed not only different development approaches, but also differences between U.S. and Japanese attitudes toward interfirm relations. "The Japanese just don't have the number of lawyers the U.S. does," explained Dick Hunder. "Trust is much more important to them. They nearly walked away when we said we wanted a clause permitting us to cancel at any time. They usually don't even spell out performance specifications in a written contract, let alone penalties."

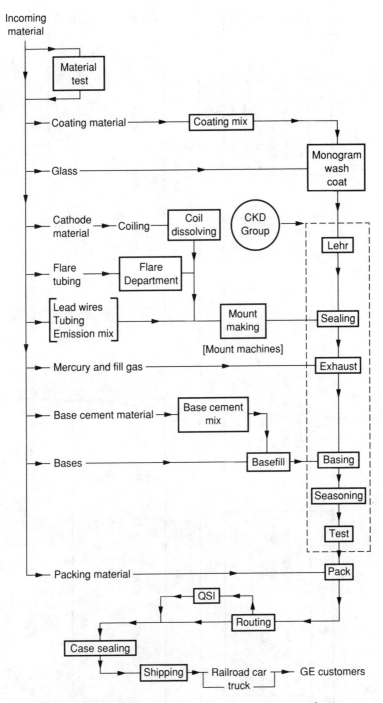

FIGURE 10.8 Basic process steps of fluorescent lamp manufacture.

Summary	
• Organized:	April 1943
• Former name:	Chukyo Elec.
• Home office:	Nagoya, Japan
• Plant location	Komaki
• Paid-in capital:	$9.7 mil.
• No. of employees:	1,376

Principal Shareholders	
• Nippon Electric Co.	19.1 %
• Matsushita Electric Co.	17.6
• Dai - Ichi Life Ins. Co.	4.4
• Sumitomo Marine & Fire Ins.	4.0
• Sumitomo Bank	3.6
• Total Shareholders	4,642

Operations	
• Automatic control eq.	16 %
• Auto. Production mach.	22
• Valves	20
• Cylinders	12
• System equipment	12
• Other	18

Earnings History
($ millions)

	1981	1982	1983	1984	1985
Sales	$156	$160	$162	$182	$226
Profit	3.63	4.11	3.71	4.00	5.56

Fiscal year: End March

Balance Sheet
($ millions)

	1981	1982	1983	1984	1985
Current assets	$64.4	$67.2	$61.4	$93.6	$105.0
Current liabilities	43.8	44.0	43.9	56.7	66.8
Stockholders' equity	39.6	32.3	44.4	54.0	74.6
Total debt	50.5	52.1	47.0	73.4	84.0

FIGURE 10.9 Background information on CKD.

had to sign a commitment that its component materials would be within agreed-upon dimensional tolerances.

Once the project was under way, CKD's fastidiousness was impressive. For example, CKD prototyped virtually everything before it committed to full-scale production equipment; GE, like most U.S. firms, tended to "cut and fit." In the early months, in fact, GE was worried that CKD would slip behind schedule because it was not yet building hardware. CKD simply responded that more planning and analysis were needed before hardware fabrication took place. CKD ended up a little ahead of schedule. On November 1, 1983, some 17 truckloads of equipment were delivered. The first good lamp came off the line a mere five weeks later.

The three months of trial and debugging activity left GE's people with high admiration for the CKD team which had studied every operating incident to correct errors as they were discovered. As the CKD team worked on meeting performance specifications, its engineers had been under very different pressures from those that Steele had felt; their work was followed at CKD headquarters in Japan, where it was evaluated in terms of the number of weeks taken to reach speed objectives. For Steele's plant people, the issue was output, since HSH was replacing the whole Jackson plant that was well into the process of shutting down. In the Lighting Leadership context, GE was very concerned about the 1984 increase in the manufacturing department's headcount, even if productivity was still rising and even if this blip in the headcount graph was clearly driven by the difficulties of juggling Jackson's closure and the HSH ramp-up.

The CKD attitude also reflected a different approach to the relationship between engineering and manufacturing. The CKD control terminal screens, for example, had space for only three head lockouts at a time, implying that the line be shut down if more than three exhaust heads were malfunctioning; GE engineers, on the other hand, would normally have been willing to continue running the line with substantially more than three of the 48 heads locked out. "They look for the facts and rely less on judgment than we do," Hunder explained. He listed other factors that he felt characterized CKD's engineering excellence—a very stable work force, "CKD ends up with particularly competent people"; components are sent out to contractors much more systematically ("since that relationship is so stable, they get the same experienced people doing the job over an extended period of time"); and "they do lots of overtime."

As MTPD had hoped, working with CKD had indeed served as an opportunity for some important in-house developments. The LSPD Engineering Department under John Breen (whose focus was product technology and basic process engineering) developed a "computerized bench" model of the HSH at LSPD headquarters in Nela Park, Cleveland. A powerful program (on a Digital Equipment VAX computer) with a very user-friendly interface and linked to a fully flexible, stripped-down HSH production line permitted LSPD Engineering to experiment using computerized control of a large number of process parameters (like pressures, gases, and temperatures) that in Bucyrus could be changed only with difficulty. Use of a bench model permitted evaluation of process designs through study of very small batches.

Perhaps the most important GE input to the HSH project was through the two major components of the process control system—the system's controller, and the Manufacturing Information Diagnostics Assistance System (M-I-D-A-S). GE wanted to avoid getting locked into the custom software control system that CKD was offering. Instead, the central group was controlled by a system based on Series 6 programmable controllers (manufactured by another GE business) with an architecture sketched out by MTPD for which CKD wrote and wired the detailed "ladder logic" (GE wired the logic for the coater). This ladder-logic approach allowed a great degree of transparency, since programmable controllers emulate the diagrams that electricians follow. This made changes more cumbersome than in a pure software system, but GE's plant people were not well-versed in programming[*]. While not a closed-loop system, these controllers allowed operators to identify and rapidly shut down any seal or exhaust head that seemed to be causing problems. They thus permitted operators to redirect into storage bins lamps coming into a problem area from upstream processes.

This open-loop control capability was to be complemented by GE's development of the M-I-D-A-S system, which, when completed, would allow detailed monitoring of, but not changes to, individual processes for quality control and diagnosis. M-I-D-A-S would allow operators to call up, on any of several terminal screens installed along the line, alert information for a given machine, yield data on any subproduct, statistical analysis and graphic display of such data, reports on specific sensors and heads, machine production and status, and overall group production information. The challenge of implementing the controllers and M-I-D-A-S for the central group and the coater was, in March 1985, still consuming most of the electronic/electrical staff's attention. Eventually, however, these systems would be extended to all the peripheral equipment and integrated for the complete line.

There had been some anxiety about the "hand-off" of the process controllers from CKD to the electrical engineer on the project, but everything had gone very well. Of course, when, in response to the inevitable minor problems, the electrical engineer began patching the system, CKD was no longer responsible for its performance. However, Bob Mitchel, the HSH project manager, had required the engineer to document the changes made so that if the person left the position or a second engineer joined the project there would be continuity.

Another focus of GE's effort was in the area of product flexibility. Its main concern was with different length lamps. The results achieved had been impressive: whereas changing the production line to allow for a lamp one foot longer, for example, had formerly taken 16 labor-hours, on the HSH line it took 2 minutes on the central group and only a total of two labor-hours for the whole line. GE had

---

[*]The HSH line manufacturing engineer and the two electricians were each given 80 hours of electronics training.

also had CKD use self-centering grippers and specially designed cradles to enable the production of lamps of different diameters. (This capability, however, was as yet untested in 1985.)

The only serious delay in the HSH process had been caused by GE. The mount machines were MTPD's responsibility, and no one had imagined that a moderate increase in their speeds would pose much of a problem. However, the combination of an absence of good documentation on the preceding generation of mount machines and an in-plant prototyping approach caused a six-month delay and required the addition of a fourth mount machine (costing nearly half a million dollars) to ensure continuity of operations. (After the installation of the first three mount machines, some 180 mechanical and 100 electrical changes were necessary.) When the CKD performance test was run, it had to be scaled back because they still were not working satisfactorily.

10.6 RUNNING THE HSH

The HSH had brought with it a change of culture. The big step forward in performance capabilities and the fact of fixed investment called for changes in pay scales for the HSH people. There was also the need for different expertise in the plant, some of which had to come from outside. Steele arranged to have CKD put a resident engineer in the plant for the first two years and hired another engineer employee who had experience with CKD's horizontal lamp-making equipment. An MTPD engineer was also assigned to the plant for the first few years.

Working on the HSH itself was a crew of 12 hourly employees ("directs") under a technical leader. The additional automation, of course, affected the traditional division of labor. HSH's continuous process technology by its very nature required three-shift, seven-day-a-week operation. The old vertical line system—two shifts, five days—had to be abandoned. The 12 directs didn't seem to mind: they welcomed the premium pay for shift work and the experience of working on the project. A special operating schedule was discussed and developed with the union, which satisfied all parties, the employees, the union, and the company. Said Steele: "If we're going to be competitive in this country, we're going to have to invest in new process technologies, and those investments will have to pay off."

The level of automation also called for considerable new training because preventive maintenance by the operators was critical to minimize downtime. These responsibilities as well as new skills needed for the new machines justified slight increases in their pay rates. Through the new training, operators became more versatile, and job classifications were reduced from 15 on the vertical line to 5 on HSH; pay rate categories were reduced from five to three (see Figure 10.10). Steele had hoped to achieve a single rate and a single classification for the whole team. While by March 1985 this had not been fully achieved, he was confident that progress was being made in that direction.

HSH staffing found most of its problems were in the area of support. It was exceedingly difficult to recruit high-quality manufacturing engineers or technical

Vertical	Job Rating	HSH	Job Rating
1 unit maintenance	(R19)	3 unit maintenance	(R19)
1 mount mechanic	(R19)	3 mount/phase maintenance	(R19)
1 unit attendant	(R14)	3 unit attendants	(R14)
1 packer	(R 7)	2 coater attendants	(R14)
1 glass handler	(R 9)	1 shrink analyst	(R 9)
1 coating maintenance	(R14)		
1 lehr mechanic	(R16)		
1 basefill	(R14)		
1 bulb loader	(R 7)		
1 mount operator	(R 7)		
1 mount loader	(R 9)		
1 base machine loader	(R 7)		
3 coater loaders	(R 7)		
1 lehr loader	(R 7)		
2 utility	(R10)		

1 rating point = 10–15 ¢/hour + OH(64%) over base pay.

FIGURE 10.10 Labor requirements on vertical and HSH units.

foremen for the night shift in Bucyrus, a small town, two hours from Cleveland. Steele believed that responsibility for a high-powered automated system like HSH would provide an ideal first step in a manufacturing management career. But there was a significant differential between the salary he could offer a graduate engineer coming out of GE's two-year Manufacturing Management Program and the salary that GE's Aircraft Engine Group could offer the same candidate for a comfortable, first-shift job in the aviation industry.

10.7 THE NEXT STEPS

All told, the successes of the HSH project were substantial, as were the problems that had been overcome. Delivered in November 1983, it had passed its performance test in late January 1984, three months ahead of schedule. After one month of hands-on training during February, the GE operating team was running HSH in regular production. By August of the same year, speeds had been increased from 4500 to 5250 units/hour and by December to about 6000 units/hour. Daily lamp production had climbed at a rate very close to target (see Figures 10.11 and 10.12).

It was, of course, of some concern that efficiency of the CKD portion of the line averaged only 79.6% in December, when the target was 90%.[*] But Steele was quite sure that the last few months' problems were a reflection of the inevitable transition bugs associated with acceleration of the line. He agreed with his staff that

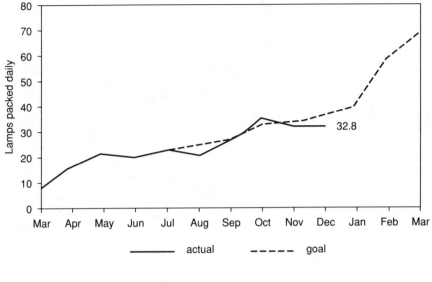

FIGURE 10.11 HSH lamp production.

on the second HSH line the goal should be 7000 units/hour. The LSPD Engineering Department had tested this rate on the computerized bench and learned that it would need an increase in the number of heads on the seal, exhaust, and aging turrets because the physical processes were pushed to their limits already. If they could be run reliably at the higher speed—CKD's challenge—then 7000 should be a feasible number.

A second HSH unit for Bucyrus was approved based on this figure; this would allow the shutdown of all its remaining vertical lines. Steele envisaged the HSH #1 fully loaded with the standard F40 lamp. On the HSH #2, he wanted to be able to produce the Watt Miser and Watt Miser Plus lamps (which had different bases and gases) and lamps with special coatings.

The real challenges regarding flexibility would come if Steele pursued his original plan of two more units for the Circleville plant. This plant would focus on the smaller volume, higher variety, fluorescent lamps. Steele decided he could

---

*This was measured independently of any limits on overall output due to the mount machines. In December 1984, mount machines efficiency was running at 61% with 12.7% loss. They were officially rated at a capacity of 3500 units/hour.

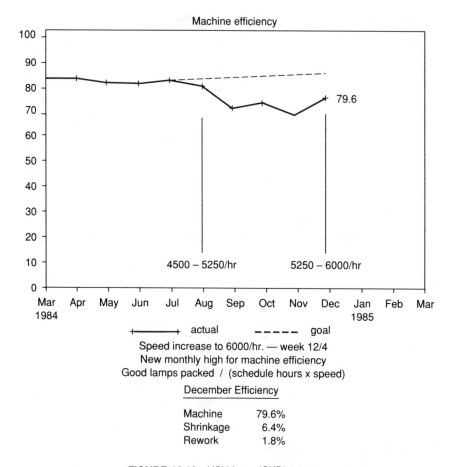

FIGURE 10.12 HSH lamp (CKD) system.

wait until then to tackle the implementation of full-scale flexibility with regard to lengths and diameters.

But all this left him with a series of questions. Even the limited flexibility he wanted to see on HSH #2 could test the Bucyrus plant's capabilities. Maintaining the fine-tuning of process parameters for efficiency at the 7000 units/hour rate would be a challenge under any circumstances, but if product specifications continually changed, perhaps the plant would need some of the engineering groups' expertise with the computerized bench.

Then there was the question of mount machines. With the increased speeds, four should have sufficed, but after the HSH #1 experience, Steele wanted and got approval for five. This was an expensive solution, but he hoped that this number would assure a smooth ramp-up. MTPD had had an engineer in residence at Bucyrus, but he'd recently left MTPD to join the Bucyrus plant staff, and the

PRO-80 initiative was still mobilizing a lot of MTPD's attention. Steele wondered whether it might be more efficient to look to an outside supplier.

He also pondered the problem of the "technical supervisors" needed for the night shifts. The technical leaders currently on HSH #1 had done a fine job. They knew lamp-making intimately (all had more than 15 years of experience), and they had learned the HSH technology through the start-up. But in the future, he felt, he would need to attract people with a different profile. Moreover, M-I-D-A-S wasn't fully operational yet, and that might make quality control difficult at higher speeds. With processes stretched to their limits and such demanding speeds in materials handling, Steele expected the plant would need a different sort of expertise. The plan had been to hire a second electrical engineer in the second quarter of 1985, but because of the delays in the implementation of the computer systems, another electrical engineer was to be hired just as soon as possible.

Later, what sort of capabilities would they really need in the Circleville plant? What sort of computer control would it take to run production at the higher speeds with the flexibility he envisaged? Would some sort of closed-loop system be necessary? Steele recalled discussions on this topic early in the planning stages for the first HSH. At that time, closed-loop control seemed unnecessary for a unit producing only one product. The speed of the system was perhaps going to be harder to maintain without it, but in any case, the plant just didn't have the kind of electronics capability that closed-loop control would require. Steele hoped these were issues he could put aside until HSH #2 was implemented.

ACKNOWLEDGMENTS

The author wishes to thank Paul S. Adler, professor of Industrial and Management Engineering, Stanford University and Professor Steven C. Wheelwright, School of Business, Harvard University for their contributions to this chapter. The material is included with permission.

QUESTIONS

10-1 Discuss the key lesson learned from the General Electric High-Speed Horizontal Project. What should be done differently? Why?

10-2 Outline the technology management ideas presented in the HSH project.

10-3 Outline the productivity improvement ideas presented in the case study.

10-4 Discuss the characteristics of the technical leaders of the project that made them successful.

11
ROLES AND RESPONSIBILITIES IN NEW PRODUCT DEVELOPMENT

11.1 INTRODUCTION

This chapter presents the roles and responsibilities of the functional organizations in product introduction. Three major product development phases will be addressed: design feasibility (phase 1), product feasibility (phase 2), and manufacturing feasibility (phase 3). Figure 11.1 illustrates the relationship between these phases and the organization's structure.

The system depicted in Figure 11.1 is a dynamic project management system, with activities and responsibilities flowing both vertically and horizontally. The design teams play a coordinating role to ensure integration and adequate transfer of the technology from one organization to another.

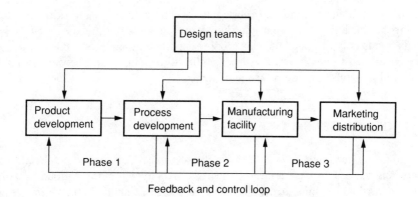

FIGURE 11.1 New product development process.

241

11.2 ROLES AND RESPONSIBILITIES

Organization of the product development effort and effective product management are challenging tasks. The challenge lies in the fact that both product development and product management are activities that cut across many departmental lines. Resources from many areas must be brought together for the successful management of evolving and existing products.

Some firms assign the responsibility for product development to the marketing department, which places a heavy emphasis on consumers and markets. However, there are dangers inherent in this assignment. The time frame of Marketing's time frame may be too short for truly innovative products; its emphasis may be on products that represent only minor improvements in existing brands; and it may neglect major technological opportunities.

Since resources from departments besides marketing are necessary to develop products, many firms have evolved the product-manager system to link the functional areas necessary for a product's success.

Although the product-manager form of organization is most commonly used for established products, some firms also assign product managers to develop new products. The advantage of this system is that the product manager is already familiar with the complexity of product management. One disadvantage is that the product manager is accustomed to being rewarded for short-term results. Also, if the new product is an extra responsibility, it may suffer from neglect. For these reasons, product-manager systems tend to produce too many product line extensions or minor modifications. Innovative product design and development projects do not fit well with the typical product manager's skills and orientation. Product-manager systems are more appropriate for incremental growth through product improvement than for breakthrough innovations.

Because of the long-term, high-risk nature of product development, many firms establish a separate department to integrate and coordinate the company's capabilities and bear the responsibilities for product innovation. This organizational form has several advantages: (1) establishing innovation as a high-priority activity, (2) balancing R&D and marketing, (3) bringing together a diverse set of skills, and (4) freeing the managing team from the short-term pressures of the existing business. Such a group may be called the new product department, and product design teams may be formed to emphasize its responsibility to develop major new innovations that build new business areas for the company.

11.3 DESIGN FEASIBILITY

Design teams frequently consist of researchers, product developers, process developers, and manufacturing engineers. The early involvement of these groups usually helps to facilitate a smooth transfer from one stage to another, thus eliminating potential problems that occur at the interfaces and speeding the new product's development cycle.

DESIGN FEASIBILITY   243

In this early stage, the design parameters are established and the design modules integrated together. Design verification testing is performed to ensure that the design conforms to the forecasted goals and constraints. The program concepts are reduced to detailed plans that respond to defined requirements in the form of performance, cost, quality (initial and early life), reliability, maintainability, and schedule objectives. Descriptions of these plans follow.

Program Plan. The purpose of this plan is to provide a comprehensive document that defines the basic user requirements, program objectives, and schedules. This document should contain the following items:

1. Program description.
2. Technology enhancements.
3. Cost:
   First user requirements.
   Objectives.
   Estimates.
4. Reliability:
   First user requirements.
   Objectives.
   Estimates.
5. Schedule and key dates.
6. Key dependencies.
7. Market forecasts.
   Users/volumes.
8. Unique user requirements.

It is the responsibility of the product program manager to develop and produce this document. Nevertheless, product development, process development, and manufacturing have a concurrent responsibility, and should be in agreement with its contents. The actual responsibilities for performing the tasks associated with the above plans cut across many departments, functions, and divisions. It is the program manager's responsibility to integrate the inputs.

Test Plan. The test plan is a design feasibility evaluation. It is the evaluation of new technologies, products, processes, or designs, and is done to identify high risk areas. Hardware is normally manufactured on a pilot line facility without full quality engineering and quality assurance support. The test plan's position is based on assessments of risk areas as identified by the testing, and on a commitment to resolve areas of concern. This document should contain the following items:

1. Technology/product description.
2. Technical concerns.

3. Test description and procedure.
4. Performance criteria.
5. Schedule and dependencies.

The test plan is performed either by reliability engineering or product assurance, depending on the organizational structure. Product development, process development, and the manufacturing facility must review and concur with the test plan and the final position.

Product Development Plan. The purpose of this plan is to provide a comprehensive document that describes the design and the program requirements. This document should contain the following items:

1. System design criteria.
2. Key dates.
3. Product description.
4. Reliability objectives.
5. Cost objectives.
6. Product maintenance/serviceability objectives.
7. Process/test description.
8. Technical concerns.
9. Support requirements, contingencies, and plans.
10. Make/buy plan (vendor sourcing).
11. Critical processes and material.
12. Development schedule.

It is the responsibility of product development to produce this plan. Nevertheless, process development and manufacturing have to concur with the plan.

Design Release Methodology. The purpose of this plan is to define, for manufacturing, the information that will be released for the product and how that information will be released, and to ensure that the product design will have adequate software support to provide the required data. This document should contain the following items:

1. List of engineering documents.
2. Methods of release.
3. List of software requirements.

It is the responsibility of product development to establish this document. However, the manufacturing facility should concur with it. This work package should be established in such a way that it ensures conformity with the data collection/acquisition systems that exist in the manufacturing facility. In an automated factory setting, computer integration techniques could be used in order to ensure compliance with all the constraints that exist in the system.

Critical Process and Material Controls. Critical process specifications are required to control product parameters that cannot be practically monitored in the manufacturing environment, and which can have an impact on product reliability. Similarly, critical materials specifications are required to ensure that all the raw materials are properly controlled in order to comply with all the quality and reliability constraints. This document should contain the following items:

1. List of critical processes and specifications.
2. List of critical materials and specifications.
3. Plans for formal release.

It is the responsibility of process development to identify the critical processes and materials, and to establish the associated specifications. Product development and the manufacturing facility have a concurrent responsibility.

Quality Plans and Requirements. The purposes of this document are (1) to define the quality requirements and procurement, and the hardware necessary to meet the commitments through phase 1; and (2) to ensure product and process compliance with quality and reliability requirements. This document should contain the following items:

1. Process routing and flow.
2. Listing of quality/inspection specifications.
3. Process/defect data to be collected.
4. New material listing and specifications.
5. Quality test requirements.
6. Quality and reliability concerns with respect to process and product.

This document is the result of a joint effort between process development and the manufacturing facility. The product development group should concur with this package. The need for such a document increases as the product gets more complex, since it serves as a tool to clarify all the interfaces and to highlight all the risk areas that need to be resolved prior to production scale-up.

Process Development Plan. The purpose of this document is to clarify and execute a process development effort that will lead to the implementation of a manufacturing system that will produce the desired product to the projected cost, quality, and reliability objectives. This document should contain the following items:

1. List of processes and alternatives.
2. Process description and alternatives.
3. Process development schedule.
4. Hardware requirements.
5. Process transfer plan.
6. Results of feasibility study.
7. Process optimization plan.
8. Equipment requirements.
9. Process routings.

It is the responsibility of process development to generate this document. Product development and the manufacturing facility have to be in complete agreement with all the areas addressed by this package. This is one of the most critical documents generated in the design feasibility stage. Cases have demonstrated that companies that have tried to bypass this stage were penalized with quality and manufacturability problems. These problems did not show up until the manufacturing scale-up phase, so they often resulted in delays in the product delivery dates. The increase in product complexity makes this stage even more important.

Process Optimization Plan. The purposes of this package are (1) to establish a plan to identify process tolerances, and (2) to establish the control parameters in order to maximize yields and minimize defects. The objective is to achieve an efficient process that is compatible with the existing manufacturing processes and requirements. This document should contain the following items:

1. Experimental matrix to determine the operating limits.
2. Schedules.
3. Process routing.
4. Equipment requirements.
5. Commitments.

It is the responsibility of process development to produce this package, but product development and manufacturing must concur with its contents. Process optimization is a critical activity of process development. Proper application of this activity could positively influence yield, quality, cost, performance, and reliability.

Manufacturing Impact Assessment Plan. The purpose of this document is to define the impact of the new technology or product on the manufacturing equipment and process. This document should contain the following items:

1. Manpower requirements.
2. Equipment impact.

3. Process requirements.
4. Material requirements.
5. Facilities impact.
6. Maintenance requirements.
7. Vendor impact.
8. Data processing requirements.

It is the responsibility of the manufacturing facility to generate this document. The document should be reviewed by the product and process development groups. This document assesses all the changes that need to be addressed, to ensure complete compatibility prior to the new product scale-up.

Automation Strategy/Plan. The purpose of this document is to define which new or existing data-processing system will be used for tracking, controlling, and analyzing the product, process, and quality information. This document should contain the following items:

1. Definition of the systems, and methods to handle yield and defect data.
2. Development support requirements.
3. Traceability and release data.
4. Machine, process, and test data.
5. Process parameter data.

It is the responsibility of the manufacturing facility to generate this document. The document should be reviewed by the product and process development groups.

11.4 PRODUCT FEASIBILITY

The program concepts from the design verification stage are reduced to detailed plans. These plans respond to defined requirements for performance, cost, quality (both initial and early life), reliability, maintainability, and schedule objectives. In this phase the blueprint designs are transformed into product prototypes. This stage of the work normally is the responsibility of product development personnel. Process development, manufacturing personnel, and others are usually active participants, however.

This phase is critical in a product development cycle. Cases have shown that companies that tried to bypass this phase have ended up with a product that either did not conform to the design parameters or did not meet the announcement dates. An important factor in this phase is the active involvement of all the team members. It is important that all the participants be in agreement with all the work packages. Typically, such an early agreement will help in reducing the product development cycle, since the parties downstream of the product can anticipate the changes required to phase the new technology into their systems. This will help in coordinating activities among product development, process development, and manufacturing, and in avoiding incompatibility of product and process designs.

Prior to starting this phase, all the documents (described in Section 11.3) should be updated to reflect all the changes and any additional requirements that are needed. Incorporation of the test results from the design feasibility stage is especially important. This updating should be completed prior to starting this phase, to provide a consolidated review that describes the design limitations and risk assessment of the product and to establish a technical position based on committed actions. This new document should contain test results from phase 1, newly identified concerns, and required actions. Additionally, the process optimization document should be completed prior to starting this phase, in order to provide a consolidated review describing the results of the process optimization plan. This package should contain the operating limits, control limits, processing limits, specification limits, process routing, and hardware definition. It is important that all the groups involved be in complete agreement with all the contents of these documents.

The program plan at this stage is implemented such that key objectives are tested against defined needs to demonstrate feasibility. The plans corresponding to this stage are as follows.

Quality Plan. The purpose of this document is to define actions that must be taken by quality engineering and quality assurance to control the quality of the product. This document contains the following items:

1. Process routing.
2. Process qualification plan.
3. Quality standards.
4. Product qualification plan.
5. Design description.
6. Product traceability requirements.
7. Equipment qualification plan.

The process development group is responsible for establishing this document. Product development and manufacturing must concur with its contents.

Process Qualification Plan. The purpose of this document is to define the necessary work and documentation required to obtain qualification for all the processes. This document should contain the following items:

1. Processes to be qualified.
2. Responsible areas.
3. Quantity of parts to be produced.

4. Schedule.
5. Process parameters to be verified.
6. Specifications required.
7. Acceptance criteria.

It is the responsibility of process development to establish this document, but product development and manufacturing must concur with all the contents of this document.

Manufacturing Scale-up Plan. The primary objective of this document is to ensure a formal shippability qualification of technology products from a qualified, engineering-controlled production facility. This document should contain all the plans that are necessary to ensure that:

1. Processes can be controlled, are repeatable, and produce a product that meets the specifications at the projected yield and defect levels through the first year of shipment.
2. Critical materials can be controlled and meet the specifications.
3. Critical process controls are adequately implemented.
4. Manufacturing tools required for the volumes committed through one year beyond the first customer shipment are adequately exercised and qualified.
5. The quality system is adequate to maintain all the required controls.
6. All the parties are committed to the reliability objectives and the improvement plans.

Manufacturing is responsible for generating this document, but product development and process development have to concur.

Process Commonality Evaluation Plan. The purpose of this document is to evaluate the processes and materials of duplicate manufacturing sources, in order to identify those items that are common (identical) and those that are different. Areas of difference should be further assessed to ensure that differences are justified. This document should contain the following items:

1. List of processes and materials by manufacturing sources.
2. Criteria for justification of differences.
3. Criteria and schedule for verifying compatibility.

It is the responsibility of manufacturing to produce this document, although product development must review it upon completion.

Phase 2 Test Plan. The purpose of this document is to evaluate the products that have been produced in a qualified and controlled production facility, under controlled-release documentation, against defined performance specifications. The hardware supplied for this test must be accompanied by a quality assurance document defining routings, processes, yield, scrap, and any deviations existing in the parts.

The test plan should include adequate sample sizes that cover the areas of technical concern and permit assessment of the following:

1. Reverification that the phase 1 risk areas have been corrected and eliminated.
2. Whether hardware produced to the qualified process parameters meets all specifications.
3. Adequacy of the quality plan.
4. Conformance to the defined reliability and performance objectives of the application.

In addition, the test plan should include:

a. Technology/product description.
b. Application conditions.
c. Test hardware description.
d. Pass/fail criteria.
e. Schedules.
f. Responsible areas.

It is the responsibility of product development to produce this document. Nevertheless, process development and the manufacturing facility must concur with all the items.

11.5 MANUFACTURING FEASIBILITY

At this stage, the program plan is implemented such that key objectives are tested against needs to demonstrate manufacturing feasibility. This stage typically starts with the successful completion of all the items in the product feasibility stage. It is important that all the results be documented and reviewed by all team members. Areas of risk should be identified and action plans should be established to resolve these areas prior to product announcement.

The program ownership is shared between process development and the production facility. To ensure compatibility between the existing manufacturing processes and the proposed process flow is critical for developing the new product. In this stage, all the processes should be optimized and qualified, and the qualification plan should demonstrate that the present manufacturing setup is capable of supporting the volume requirements through the early stages of the product's life cycle. Also at this stage, all the design feasibility and product feasibility plans should be updated and reviewed in order to demonstrate compliance with the program objectives as identified in the program plan. The impact of any deviation should be assessed; measurements in terms of reliability, cost, time delay, and so on, should be reviewed; and plans should be put in place to achieve compliance. Following are some of the important activities that must be completed during this phase:

1. Perform all the process qualifications, in order to ensure that the manufacturing processes will be capable of meeting the projected performance.
2. Ensure that the achieved reliability is in compliance with the targeted reliability growth. Improvement plans should be established and committed if the achieved reliability falls below the lower confidence limit of the targeted reliability.
3. Ensure that the product meets the requirements of the Manufacturing Scale-up Plan established in the product feasibility phase.
4. Provide a methodology for the control of product, process, and equipment changes.
5. Define the method and responsibility for obtaining detailed knowledge of the performance of the new product. This is of major importance for tracking and calculating failure rates and for determining if it is necessary to introduce further reliability improvement programs.

Several factors are important for the successful demonstration of manufacturing feasibility. Successful implementation results from the combined effort of the functions within and across organizational boundaries. Additionally, many implementation attempts fail or are only marginally successful because the process design as it is ultimately installed and integrated into the work flow of the organization is incorrectly matched to the product needs of the user organization. This disparity results from the lack of early involvement by the product and process development teams.

Thus, it is important for all groups to participate in all phases of the new product development process; such participation will lead to lower product development cost, shorter product development cycles, and a reduced number of engineering changes. This has been demonstrated by the Japanese industry's use of "quality function deployment," whereby the marketing people, design engineers, and manufacturing staff work closely together from the time a product is first conceived. This early involvement often enhances the communication between all the involved parties, and resulting in smooth product transfer.

11.6 DESIGN/MANUFACTURING/MARKETING INTERFACES

The transfer of technology from design to manufacturing and then to marketing is a complex process. The complexity is further increased by the existence of barriers at the interfaces. These barriers may have a negative impact on the new product's development and success.

252 ROLES AND RESPONSIBILITIES IN NEW PRODUCT DEVELOPMENT

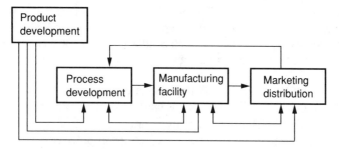

FIGURE 11.2 Product development interfaces.

The effective transfer of technology across the interfaces has a direct bearing on the success of commercializing new technologies. There is theoretical and empirical evidence that conflict occurring at the interfaces is connected with incompatibility of operating or departmental goals. This conflict between R&D and manufacturing is due to the fact that manufacturing's goal orientation tends to focus on promoting and optimizing volume of output and production rates, whereas R&D's goal orientation focuses on the introduction of new products. New products will, at least initially, tend to slow production rates. Thus, the attainment of one goal seems to preclude or substantially hinder the attainment of the other.

Several interfaces exist throughout the product development cycle. The smooth transfer of technology across these interfaces is a key requirement for the product's success. Although some of the barriers are universal to all the interfaces, some barriers are unique to some interfaces. Thus, it is very important to identify these barriers in order to find ways of avoiding or resolving them, and subsequently improving the probability of success for the new product.

Figure 11.2 depicts the interfaces along the product development cycle between product development and market entry. The number of interfaces is calculated according to the following equation:

$$\text{Number of interfaces} = \frac{I!}{(2!)(I-2)!}$$

where I denotes the number of product development functions. Thus, there are a total of six interfaces between the four functions shown in Figure 11.2.

One of the key elements of a good transfer is early formation and involvement of design teams that manage the implementation of the project. These design teams, as shown in Figure 11.1, are formed from researchers, product developers, process developers, and manufacturing and marketing personnel. The early involvement of these groups usually helps to facilitate a smooth transfer from one stage to another. Souder (1987) has shown that this helps to eliminate potential problems and speed the new product development cycle.

11.7 IMPLEMENTATION

The implementation phase of the project should begin before the new manufacturing process and its equipment arrive in the production shop. The involvement of the manufacturing departments should begin before the development project is completed, and the development project staff's involvement must continue until the production implementation is completed.

The manager of the implementation phase will be responsible for transferring the technology across the departmental interfaces. He or she must therefore have authority across that interface because responsibility cannot exist without authority. The early selection, designation, and granting of authority to the person who will manage the implementation is extremely important, so that he or she can take all the steps necessary to ensure the prompt and efficient production start-up of the new technology. The assignment of design and manufacturing personnel to the new product team with the new product's staff will result in a team effort, and will prevent one of the most common causes of implementation failures.

SUMMARY

This chapter has presented the key roles and responsibilities that are involved in the following phases of a new product development:

1. Design verification (phase 1).
2. Product feasibility (phase 2).
3. Manufacturing feasibility (phase 3).

The objective of each development plan, for each of the phases, is also discussed. Additionally, this chapter addresses the barriers that exist at the interfaces between design, manufacturing, and marketing. These barriers, if not handled properly, could have a negative impact on the new product development and its success.

ACKNOWLEDGMENT

This chapter was contributed by Dr. Suheil M. Nassar, IBM Corporation, and Dr. William E. Souder, University of Pittsburgh. The material was included with permission.

QUESTIONS

11-1 Discuss the differences, in roles and responsibilities, between the design feasibility phase and the product feasibility phase.

11-2 You are a program manager in a computer company. Your responsibility is to establish a program plan for developing a new personal computer that is superior in performance, cost, and reliability to any existing personal computer on the market.

   a. Discuss the marketing/competitive analysis data required to develop the program plan.

   b. Using the above information, develop a program plan for this new product.

11-3 Discuss the differences between the phase 1 and phase 2 test plans. What are the objectives of these plans?

11-4 Discuss the disadvantages of not having an automation strategy defined during the design feasibility phase.

11-5 The role of the design team is to integrate the efforts of the various functional organizations. Discuss the skills that a typical design team should have.

11-6 What are the contributions that the early involvement of the manufacturing team has towards the success of a new product development?

11-7 Product development personnel often have a different skill base than marketing personnel. What are some of the problems that might occur at this interface?

11-8 What role should the design team play in order to reduce the conflict at the functional interfaces?

11-9 Compare the similarities between the organizational structure, as presented in Figure 11.1, and a typical matrix organization.

11-10 How would the organization structure, as discussed in this chapter, contribute to the successful implementation of a new product?

CONCLUSION

There is a close relationship between technological innovation—the process through which new technology is created—and the management of the use of technology in the workplace. Technology—as tools, methods, or knowledge—is an essential requirement for most organizations in the process of transforming resources to useful output. Technological innovation is indispensable for growth, productivity, quality improvement, and survival in a competitive world economy. Most of the success obtained by successful firms can be attributed to the design, development, and implementation of policies and programs that promote technological innovation.

This book has provided significant and useful tools, techniques, concepts, and methodologies for managing the interactive, integrated process of innovation and technology issues. Technology and concepts involved in the management of technological innovation have been presented with models for managing innovation and technology. Detailed discussions on managing the process of technological innovation have been offered. New ideas have been proposed to guide decision makers in formulating technology and innovation policies at the firm level.

A comprehensive design rule and framework have been presented to aid system designers, developers, and users of technology. Several new techniques and methodologies for justifying and implementing new technologies have been discussed, as have implementation strategy for new technology in the workplace and common problems experienced.

A framework for managing the utilization, transfer, and forecasting of technology was also presented. Specific guidelines for designing a product for automation and production have been discussed, along with various concepts in automation application. The value engineering concept has been presented as a technique for reducing the cost of parts and products in the design phase. The product life cycle and the ideal manufacturing goals have been discussed, along with other factors affecting product design for automation. A solution strategy has been provided to aid system designers in minimizing the impact of product design on the manufacturing processes.

The concept of group technology has been explained, and the factors that can and cannot be controlled by management in research and development organizations have been discussed. The characteristics of research and development organization, an organizational and leadership model for the R&D environment, and the role of the innovative manager have been examined.

Current technologies such as robotics, computers, and knowledge-based expert systems design rules and how to implement them have been discussed, and a guideline for successful supervisory performance in a research and development environment has been presented. At the end of each chapter, questions and problems are presented to stimulate the reader's experience, suggest scope for further investigation, or use as exercises for students and workshop participants.

Perhaps the materials provided have broader use and applicability because they have been validated in real-life situations. Case studies from major companies, such as IBM and General Electric, and from successful medium-sized companies are presented to illustrate the sensitivity of the various techniques to the real world. Where applicable, self-assessment checklists have also been provided to aid decision makers in high-technology implementation such as computer-aided manufacturing, robotics, expert systems, and computer-integrated manufacturing.

Perhaps the three most important factors that promote technological innovation are (1) an effective system of management of people and resources; (2) an environment that encourages entrepreneurial initiative at all levels of the organization; and (3) the ability of individuals within an organization to generate resources, take risks, and commit to a progressive course of action. Other factors such as adequate training for the work force, management orientation, project selection criteria, organizational structure, government influence, and economic climate can have impact on the rate of technological innovation. The management of technological innovation requires a flexed contingency leadership style to accommodate the complex interactive processes involved in the innovative environment.

One asset that an organization must have in a technologically innovative environment is strong planning ability. Proper strategic, tactical, and operational planning must be in place to provide the organization with the ability to respond to the dynamic changes in technology life cycles, people, processes, markets, and organization and competitive factors. Equally important are the tools, techniques, and programs provided for the ongoing improvement in the innovation and technology management process.

Although the introduction of various technologies in the workplace has led to significant gains in productivity and improvement in the quality of working life, an ongoing effort must still be made to manage the potential negative impact of technology. There is no doubt that human resources are the greatest asset for any organization. Without human production of new ideas, the development and use of technology will not happen. An ongoing effort must therefore be expended to minimize the negative impact of technology on human beings.

In the next several decades the quality of working life will continue to depend on how well technology impact issues are addressed. The need exists for the following actions.

- Ongoing cooperative efforts among industry, academia, and government in formulating and implementing the right technological policies at the international, national, and firm levels.
- Ongoing cooperative efforts among the designers, developers, and users of technology in addressing technology-based problems and issues.
- Analyzing and understanding the task environment thoroughly before introducing a new technology.
- Ongoing management system in place for managing potential problems that the change to a new technology may bring to task content, interaction processes, workers' beliefs, and the quality of working life.
- Use of technology-based systems as mechanisms for improvement in the production and service processes and not as a means to control workers.
- Provisions in technology-based systems to advise, alert, or warn users of potential events. This further implies that designers must find a compromise between those aspects of the system that influence system productivity and quality and those that affect the quality of working life.
- Making technology-based systems that are compatible with (1) the practice, culture, and job satisfaction of workers; and (2) the structure of the organization.
- Designing and implementing policies that provide a good fit between innovative individuals and people in general, tasks, and the total organization.
- Designing and implementing new training programs to assist workers affected by dislocation due to technology; implementation of retraining opportunities for workers who have to change careers because of new technology.
- Providing adequate procedures for long-range planning before implementation of new technology.
- Providing advance notice and consultation to employees about impending changes.
- Providing adequate procedure and communication channels for handling grievances on seniority rights, financial requirements, quality of working life issues, and other negative impacts of the new technology.

Innovation and technology management is a dynamic field. The need therefore exists for ongoing investigation in the following areas.

- Methodologies and techniques for justifying new technologies that improve productivity, product quality, and the quality of working life.

- How to evaluate the impact of existing and new technologies on performance and human health.
- How to evaluate the effects of technological change on the skill requirements of the workforce and the social consequences of new technology.
- Adequate procedures and methodologies for matching and training of the existing skilled workforce to the requirements of new technologies.
- Techniques for project portfolio selection and management in high-technology environments.
- Factors leading to a reorganization of technological activity in firms.
- Advantages and disadvantages of different organizational structures on the organization of the product development cycle, from concept to production.
- How to identify the facilitators and inhibitors of technological innovation within organizations.
- How to integrate technological and strategic plans.
- The impact of legislation, regulation, and insurance on technological change.
- How to increase the user's influence in the development, selection, and application of new technologies.
- How to implement new technology in the work environment and minimize resistance to change.
- Appropriate gain-sharing techniques for distributing the benefits from new technology.
- How to assess the effectiveness of research and development units and activities.
- How to manage dynamic changes in the technology life cycle.
- New methodologies for technology feasibility studies, specification monitoring, model development, prototype testing, and product manufacturability and survivability.
- How to provide full systems ability for technologies to communicate with each other under one system network.
- Methodologies and techniques for managing technology asset information and high-technology procedures and languages.
- New motivation techniques that address the changing values and roles of workers as technology changes are needed.

Managing innovation and technology in a competitive environment requires the attention of business managers and decision makers to (1) provide adequate facilities for human resource development under conditions of rapid technological and social change; (2) provide ongoing support for research and development activities through appropriate funding and project management; (3) implement appropriate strategies and controls to monitor the interfaces between the operating organization and the external environment; (4) provide the right methods and tools for managing technical resources and organizational complexities caused by technological changes; (5) provide the right synergy between the rate of innovation, product manufacturability, and marketability; and (6) continuously evaluate the positive and negative consequences of technology in the work environment and implement corrective action when necessary.

It is clear that the ability of a firm or enterprise unit to survive in a competitive environment depends heavily on the degree to which it can implement new technological breakthroughs to improve productivity, quality, efficiency, and overall effectiveness. The success of an organization in managing innovation and technology depends on the skills and vision of the employees and managers who are charged with the responsibilities of running the various production and service processes. Teamwork among employees and managers; suppliers and producers; and designers, implementers, and users will improve an organization's effectiveness in managing technological innovation.

This book has attempted to integrate innovation and technology management for the better understanding of decision makers, practitioners, students, and researchers. I hope the content presented here will provide useful insight and guidelines for both existing and potential innovators. Strong entrepreneurial, creative skills are behind many successful innovations, but most new ideas come simply from the application of common sense. The great innovator is that person who not only has lots of ideas, but also works through obstacles to bring ideas to fruition.

Innovation and technology management spans a variety of issues at international, national, and enterprise levels. It therefore requires a systems approach for understanding the impacts at the different levels. Competing in the world market will continue to depend heavily on the successful management of innovation and technology; that is, ensuring that all resources are optimized to efficiently manage simple- and high-technology systems.

REFERENCES

Abernathy, William J. (1982). "Competitive Decline in U.S. Innovation: The Management Factor," *Research Management*, September, pp. 34–41.

Abernathy, William J., and K. B. Clark (1985). "Mapping the Winds of Creative Destruction." *Research Policy*, 14(1), pp. 2–21.

Abramovitz, M. (1956). "Resource and Output Trends in the United States Since 1870," *American Economic Association Papers*.

Adams, Russ (1986). "The World of Automatic Identification," *Bar Code News*, **6** (3), pp. 6–18.

Allen, T. J. (1966). "Managing the Flow of Scientific and Technological Information," unpublished doctoral dissertation, M.I.T., Cambridge.

Allen, T. J. (1968). *Managing Communication in Research and Development Organizations*, M.I.T. Press, Cambridge.

Allen, Thomas J. (1978). *Managing the Flow of Technology*, M.I.T. Press, Cambridge.

Ayoub, N. N. (1966). "Biomechanical Considerations in Industrial Work." Paper presented at the Human Factors Conference of the American Society of Mechanical Engineers, ASME publication 66-HUF-8.

Ayres, Robert U. (1988). "Future Trends in Factory Automation," *Manufacturing Review*, **1**, (2).

Bartimo, Jim (1984). "Smalltalk" with Alan Kay, *InfoWorld* June 11, pp. 58-61.

Bechofer, Frank (1973). "The Relationship Between Technology and Shop-Floor Behavior." In D. O. Edge and J. N. Wolfe, Eds., *Meaning and Control*, Tavistock, London.

Beer, Stafford (1966). *Decision and Control*, John Wiley and Sons, New York.

Berger, Peter (1964). "Some General Observations on the Problems of Work." In Peter Berger, Ed., *The Human Shape of Work*, Macmillan, New York.

Betz, Frederick L. (1987). *Managing Technology*, Prentice-Hall, England Cliffs, NJ.

Betz, Frederick L. Vaughn Blankenship, Carlos Kruytbosch, and Richard Mason. *Allocating*

R&D Resources in the Public Sector, in Management of Research and Innovation. B. V. Dean and J. L. Goldhar, eds. New York: North-Holland, 1980.

Bitondo, Domenic, and Alan Frohman (1981). "Linking Technological and Business Planning," *Research Management,* **24** (6), pp. 19–23.

Bjorn-Anderson, Niels (1976). "Organizational Aspects of Systems Design," *Data* **12**, pp. 75–80.

Blauner, Robert (1964). *Alienation and Freedom: The Factory Worker and His Industry,* University of Chicago Press, Chicago.

Bok, Derek (1964). *Automation, Productivity, and Manpower Problems,* President's Advisory Committee on Labor-Management Policy, U.S. Government Printing Office, Washington, D.C.

Booz, Allen and Hamilton Inc. (1982). *New Product Management for the 1980s,* Special report.

Bradley, Gunilla (1977). "Computerization and Some Psychosocial Factors in the Work Environment." In *Managing Job Stress,* H.E.W., U.S. Government Documentation Center, Washington.

Buffa, E. S. (1984). *Meeting the Competitive Challenge,* Homewood, Illinois, Dow Jones-Irwin.

Burbridge, J. L. (1975). *The Introduction of Group Technology,* John Wiley & Sons, New York.

Business Week (1984). "Apple Computer's Counterattack Against IBM," January 16, pp. 78–82.

Carrasco, H., and K. Kengskool, (1988). "Justification Methodology for New Technology." *Proceedings for the First International Conference on Technology Management,* Miami.

Cho, Ka Kyu and I. Ham (1985). "Group Scheduling Heuristics for Minimum Makespan in Multi-Stage Manufacturing Systems." *Annual International Industrial Engineering Conference Proceedings,* Chicago.

Chorafas, D. (1987). *Engineering Productivity Through CAD/AM,* Butterworth, London.

Clark, K., R. H. Hayes, and C. Lorenz, Eds. (1985). *"The Uneasy Alliance: Managing the Productivity Technology Dilemma,"* Boston, *Harvard Business Review School Press.*

Cohen, Stephen S., and John Zysman (1987). *Manufacturing Matters: The Myth of the Post-Industrial Society,* Basic Books, New York.

Colding, Bertil, (1977). *Delphi Forecasts of Manufacturing Technology,* Society of Manufacturing Engineers, University of Michigan; I.F.S. Publications, Bedford, England.

Corder, G. C. (1971). *Organizing Maintenance,* British Institute of Maintenance, London.

Croft, F. M., and W. N. Ledbetter (1988). "Project Management and Implementation." In *Proceedings for the First International Conference on Technology Management,* Interscience Publishing Company, Geneva, Switzerland.

Daniels, G. S. (1958). "Human Engineering Investigations—Design Guides," Convair Division of General Dynamics Corporation Report No. HFE-64-20.

Daniels, G. S., et al. (1953). "Anthropometry of WAF Basic Trainees," U.S. Air Force, Wright Air Development Center, Technical Report No. 53-12.

Danziger, James N., and William H. Dutton (1977). "Computers as an Innovation in American Local Governments," *CACM* **20** (12): pp. 945–956.

Davis, Randall (1982). *Knowledge-Based Systems in Artificial Intelligence,* McGraw-Hill, New York.

Dearden, J. (1987) "Measuring Profit Center Managers," *Harvard Business Review*, September–October, pp. 84–88.

Dooley, Arch R. , and Thomas M. Strout (1971). "Rise of the Blue-Collar Computer," *Harvard Business Review*, July–August.

Drucker, P. (1969). *The Age of Discontinuity*, Harper & Row, New York.

Drucker, Peter, (1974). *Management—Tasks, Responsibilities, Practices*, Harper & Row, New York.

Drucker, Peter F. (1985). "The Discipline of Innovation," *Harvard Business Review*, May-June, pp. 67–72.

Edosomwan, Johnson A., (1985a). "A Methodology for Assessing the Impact of Computer Technology on Productivity, Production Quality, Job Satisfaction, and Psychological Stress in a Specific Assembly Task,"unpublished doctoral dissertation, The George Washington University Department of Engineering Administration, Washington, DC. Research sponsored by the Social Science Research Council (U.S. Department of Labor) Grant Number SS-36-83-21 and the IBM Corporation Grant Number IBM-2J2-2K5-722271-83-85.

Edosomwan, Johnson A. (1985b). "Computer-Aided Manufacturing Impact on Productivity, Production Quality, Job Satisfaction and Psychological Stress." *Fall Industrial Engineering International Conference Proceedings*, Chicago.

Edosomwan, Johnson A. (1985c). "Computer Impact on Task Structure in a Production Environment. " IBM Technical Paper Data Systems Division, New York.

Edosomwan, Johnson A. (1985d). "A Task-Oriented Total Productivity Measurement Model for Electronic Printed Circuit Board Assembly." *First International Electronics Assembly Conference Proceedings*, Santa Clara, CA, October 7–9.

Edosomwan, Johnson A. (1986a). "A Comprehensive Methodology for Productivity Planning." *Industrial Engineering*, January, pp. 64–69.

Edosomwan, Johnson A. (1986b). "Managing Technology in the Workplace: A Challenge for Industrial Engineers." *Industrial Engineering*, February, pp. 14–18.

Edosomwan, Johnson A. (1986c). "Technology Impact on the Quality of Working Life—A Challenge for Engineering Managers in the Year 2000." *Proceedings for the First International Conference on Engineering Management*, Arlington, VA.

Edosomwan, Johnson A. (1986d). "Productivity Management in Computer-Aided Manufacturing Environment." *Proceedings for the First International Conference on Engineering Management*, Arlington, VA.

Edosomwan, Johnson A. (1986e). "The Role of Ergonomics in Improving Productivity and Job Satisfaction in Computer-Aided Assembly Tasks." *Proceedings for the Annual International Institute of Industrial Engineers Conference*, Dallas.

Edosomwan, Johnson A. (1986f). "A Methodology for Assessing the Impact of Robotics on Total Productivity in an Assembly Task." *Proceedings for the International Institute of Industrial Conference*, Dallas.

Edosomwan, Johnson A. (1987a). *Integrating Productivity and Quality Management*. Marcel Dekker Inc., New York.

Edosomwan, Johnson A. (1987b). "Computer Role in Decision Making in the Year 2000." *Proceedings for the 9th Annual Computers and Industrial Engineering Conference*, Atlanta.

Edosomwan, Johnson A. (1987c). "A Technology-Oriented Total Productivity Measurement Model" Productivity Management Frontier I. Elsevier The Netherlands.

Edosomwan, Johnson A. (1987d). "Productivity Measurement in Group Technology Production Environment." *Proceedings for the World Productivity Forum,* Washington, DC, May.

Edosomwan, Johnson A. (1987e). "Reducing Computer Operating Costs," *Proceedings for the 9th Annual Computers and Industrial Engineering Conference,* Atlanta.

Edosomwan, J. A. , F. Betz, G. Gaynor, and T. Khalil (1988). "Managing Technology Direction for the Future." *Proceedings for the Industrial Engineering System Integration Conference,* St. Louis.

Eldin, Hamed Kamal (1988). "Problems of Technology Transfer to Developing Countries." *Management of Technology, Vol. 1,* In J. A. Edosomwan and T. M. Khalil, Eds., Interscience, Geneva, Switzerland.

Ettlie, J. E. (1973). "Technological Transfer from Innovators to Users," *Industrial Engineering, 5.*

Faunce, William A. (1968). *Problems of an Industrial Society,* McGraw-Hill, New York.

Fergenbaum, Edward A., and Pamela McCorduck (1983). *The Fifth Generation: Artificial Intelligence and Japan's Computer Challenge To The World,* Addison-Wesley, Reading, MA.

Fichtelman, Michael (1985). "The Expert Mechanic,"*Byte,* Vol. N06 (June).

Foster, R. (1986). *Innovation: The Attacker's Advantage,* Summit Books, New York.

Foster, Richard N., Lawrence H. Linden, Roger L. Whiteley, and Alan M. Kantrow (1985). "Improving the Return of R&D II," *Research Management,* **28** (2), pp. 13–22.

Frohman, Alan L. (1981). "Technology as a Competitive Weapon," *Harvard Business Review,* March-April pp. 117–126.

Frohman, Alan L. (1982). "Technology as a Competitive Weapon," *Harvard Business Review,* January–February, pp. 97–104.

Galbraith, Jay (1973). *Designing Complex Organizations,* Addison-Wesley, Reading, MA.

Gaynor, G. H. (1988). "Selecting, Monitoring and Terminating Projects." In *Proceedings for the First International Conference on Technology Management,* Interscience Publishing Company, Geneva, Switzerland.

Gibson, J., J. Ivancevich, and J. Donnelly (1979). *Organizations,* 3rd edition, Business Publications, 1979.

Ginn, M. E. , and A. H. Rubenstein, (1986). "The R&D/Production Interface: A Case Study of New Product Commercialization," *Journal of Production Innovative Management* **3**.

Goldhar, Joel D., and Avak Avakian (1979). *Managing Change in Manufacturing Systems,* Discussion paper presented by the Industrial Technology Delegation from the United States to the Republic of China, Peking, November 26–December 10.

Ham, I. (1976). "Introduction of Group Technology," Society of Manufacturing Engineers Technical Paper MMr-76-03.

Hauser, J. R. and D. Clausing (1988). "The House of Quality," *Harvard Business Review,* May–June.

Heinrich, H. W. , D. Petersen, and N. Ross (1980). *Industrial Accident Prevention,* McGraw-Hill, New York.

Hoos, I., (1961). *Automation in the Office,* Public Affairs Press, Washington, DC.

Horwitch, M., Ed. (1986). *Technology in the Modern Corporation: A Strategic Perspective,* Penguin Press, Elmsford, N.Y.

Hunter, Stephen (1985). "Cost Justification: The Overhead Dilemma." *Society of Manufacturing Engineers Robots 9 Conference Proceedings.*

Jackson, B. B., and B. P. Shapiro (1979). "New Ways to Make Product Line Decisions," *Harvard Business Review,* May–June (1979).

Jahn, Edwin C. (1973). *Future Technological Needs and Trends of the Paper Industry: Education Implications for the University.* Technical Association of the Pulp and Paper Industry, Atlanta.

Kanter, Rosabeth Moss (1982). "The Middle Manager as Innovator," *Harvard Business Review,* July-August, pp. 95–105.

Kaplan, R. S. (1986). "Must CIM Be Justified by Faith Alone?" *Harvard Business Review,* March-April, pp. 87–93.

Karasek, R. A., Jr., and J. A. Turner (1981). Software Ergonomics: *"Effects of Computer Application Design Parameters on Operator Task Performance and Health."* Paper presented at the 1981 Fall National I.I.E. Industrial Engineering Conference, Washington, DC, December 8.

Katz, D. and R. Kahn (1966). *The Social Psychology of Organizations,* John Wiley & Sons, New York.

Kendrick, J. W. (1961). "Productivity Trends in the United States," National Bureau of Economic Research, Princeton University Press.

Khalil, Tarek (1976). "Design Tools and Machines to Fit the Man," *Industrial Engineering,* **a** (b), pp. xx–xx.

Kheel, Theodore W. (1966). "Changing Patterns of Collective Bargaining in the U.S." In Jack Stieber, Ed., *Employment Problems of Automation and Advanced Technology,* St. Martin's Press, New York.

Kling, Rob (1978). "The Impacts of Computing on the Work of Managers, and Clerks," Organization, University of California Press at Irvine.

Kunz, Stephen (1971). "Fitting the Job to the Man," *Industrial Engineering,* **3** (1).

Kuriloff, A. H., and J. M. Hemphill, Jr. (1978). *How to Start Your Own Business and Succeed.* McGraw-Hill, New York.

Lawrence, P. and J. Lorsch (1967). *Organizations and Environment,* Harvard University Press, Cambridge.

Lee, Hak Chong (1965a). "Electronic Data Processing and Employee Perception of Changes in Work Skill Requirements and Work Characteristics," *Personnel Journal* **44** (7), pp. 49–53.

Lee, Hak Chong (1965b). "Do Workers Really Want Flexibility on the Job," *Personnel* **42,** pp. 74–77.

Lenz, R. C., Jr. (1968). "Forecast of Exploding Technologies by Trend Extrapolation." In J. R. Bright, Ed., Technological Forecasting for Industry and Government, Prentice-Hall, Englewood Cliffs, NJ, pp. 65–69.

Lewis, Hordan D. (1982). "Technology, Enterprise and American Economic Growth," *Science,* **215,** March 5.

Likert, R. (1967). *The Human Organization,* McGraw-Hill, New York.

McCormick, Ernest J. (1976). *Human Factors in Engineering and Design,* 4th ed., McGraw-Hill, New York.

McElroy, J. (1987). "The House of Quality: For Whom Are We Building Cars?" *Automotive Industries,* June.

Madique, M. A. , and P. Patch (1978). "Corporate Strategy and Technological Policy" *Harvard Business Review,* May-June.

Madique, Modesto A., and Billie Jo Zirger (1984). "A Study of Success and Failure in Product Innovation: The Case of the U.S. Electronics Industry," *IEEE Transactions on Engineering Management,* **EM-31** (4), pp. 192–202.

Mann, Floyd C., and Lawrence K. Williams (1960). "Observations on the Dynamics of a Change to Electronic Data Processing Equipment," *ASQ* **5,** pp. 217–276.

Marquis, Donald G. (1969). "The Anatomy of Successful Innovations," *Innovation,* November. Reprinted in M. L. Tushman and W. L. Moore, Eds. , *Readings in the Management of Innovation,* Marshfield, MA, 1982.

Meredith, J. R. , and S. G. Green (1988). "Managing the Introduction of Advanced Manufacturing Technologies," *Manufacturing Review,* **1** (2).

Miller, Eric J., and David J. Armstrong (1966). "The Influence of Advanced Technology on the Structure of Management Organization." In Jack Stieber, Ed., *Employment Problems of Automation and Advanced Technology,* St. Martin's Press, New York.

Mogaveor, L. N. and R. S. Shane (1982). *What Every Engineer Should Know About Technology Transfer and Innovation,* Marcel Dekker Inc., New York.

Monger, R. F. (1988). *Mastering Technology: A Management Framework for Getting Results,* New York, The Free Press.

Moranian, Thomas (1963). *The Research and Development Engineer as Manager,* Holt, Rinehart and Winston, New York.

Morris, W. T. (1977). *Decision Analysis,* Grind, Inc., Columbus, OH.

Morrison, Ann M. (1984). "Apple Bites Back," *Fortune,* Feb. 20, pp. 86–100.

National Research Council (1987). *Management of Technology: The Hidden Competitive Advantage,* National Academy Press, Washington D.C.

New York Stock Exchange (1984). *U.S. International Competitiveness: Perception and Reality,* New York.

Nassar, S. M. (1988). "A Reliability/Cost Trade-off Model for Managing New Product Development," (1988). Ph.D. dissertation, University of Pittsburgh.

Norman, Colin (1980). "Microelectronics at Work: Productivity and Jobs in the World Economy," *World Watch Paper* **39**, pp. 23–25.

Ohmae, K. (1985). *Triad Power: The Coming Shape of Global Competition,* The Free Press, New York.

Peters, T. (1987). *Thriving on Chaos: Handbook for a Management Revolution,* Alfred A. Knopf, New York.

Porter, M. E. (1980). *Competitive Strategy: Techniques for Analyzing Industries and Competitors,* The Free Press, New York.

Porter, Michael E. (1985). *Competitive Advantage,* Free Press, New York.

Quinn, J. B. (1985). "Managing Innovation: Controlled Chaoes," *Harvard Business Review,* **63** (3) pp. 35–43.

Ramo, Simon (1980). *The Management of Innovative Technological Corporations,* John Wiley & Sons, New York.

Rich, Elaine (1983). *Artificial Intelligence,* McGraw-Hill, New York.

Riggs, H. E. (1983). *Managing High-Technology Companies,* Lifetime Learning Publications, Belmont, CA.

Roberts, E. B., and A. R. Fusfield (1981). "Staffing the Innovative Technology-based Organization," *Sloan Management Review,* **22** (3), pp. 19–34.

Rodriguez, Ramon, and Oscar Adaniya, (1985). "Group Technology Cell Allocation." *Annual International Industrial Engineering Conference Proceedings,* xxxxx.

Rogers, Everett M. (1983). *Diffusion of Innovations,* Free Press, New York.

Rybczynski, W. (1983). *Taming the Tiger: The Struggle to Control Technology,* The Viking Press, New York.

Saad, H. G. (1988). "An Optimization Approach for Investment Allocations to Different Technologies. " *Proceedings for the First International Conference on Technology Management,* Miami.

Schmookler, Jacob (1966). *Invention and Economic Growth,* Harvard University Press, Cambridge, MA.

Schonberger, R. J. (1987). "Frugal Manufacturing," *Harvard Business Review,* September-October, pp. 95–100.

Simon, Herbert. (1960). *The New Science of Management Decision,* Harper & Row, New York.

Simpson, J. A. (1984), "Investment Justification of Robotics Technology in Aerospace Manufacturing," Report from Applied Concepts Corp. Prepared for the Air Force Business Research Management Center, Contract No. F33615-83-C-5080.

Skinner, Wickham (1978). *Manufacturing in the Corporate Strategy.* New York: John Wiley & Sons, New York.

Skinner, Wickham (1979). "The Impact of Changing Technology on the Working Environment." In Clark Kerr and Jerome M. Rosow, Eds., *Work in America: The Decade Ahead,* Work in America Institute/Van Nostrand Reinhold Series, Van Nostrand Reinhold, New York.

Skinner, W., and K. Chakraborty (1982). *The Impact of New Technology: People and Organizations in Manufacturing and Allied Industries,* Pergamon Press, New York.

Souder, W. E., "Promoting an Effective R&D/Marketing Interface," *Research Management,* **23** (4) July.

Souder, W. E. (1987). *Managing New Product Innovations,* Lexington Books, Lexington, MA.

Sumanth, D. J. (1984). *Productivity Engineering and Management,* McGraw-Hill, New York.

Sumanth, D. J. (1987). *The Comprehensive Total Productivity Model (CTPM),* Working Document (Version 1).

Sumanth, D. J. (1988). "Challenges and Opportunities in Managing Technology Discontinuities on S-Curves," *Proceedings for the First International Conference on Technology Management,* Interscience Publishers, Geneva.

Technology Atlas Team (1987). *Technology Atlas Project,* United Nations Asian and Pacific Centre for Transfer of Technology, Bangalore, India.

Tornatzky, L. G., J. D. Eveland, M. G. Boylan, W. A. Hetzner, E. C. Johnson, D. Roitman, and J. Schneider (1983). *The Process of Technological Innovation,* National Science Foundation, Washington, D.C.

Turner, Jon A. (1980). "Computers in Bank Clerical Functions: Implication for Productivity and the Quality of Working Life," Unpublished doctoral dissertation, Columbia University, New York.

Tushman, M. (1976). "Communication in Research and Development Laboratories: An Information Processing Approach," unpublished doctoral dissertation, M.I.T., Cambridge.

Twiss, Brian (1980). *Managing Technological Innovation,* 2nd ed., Longmans, London.

Urban, G., J. Hauser, and N. Dholakia (1987). *Essentials of New Product Management,* Prentice-Hall, New York.

Uttal, Bro (1981). "Xerox Xooms toward the Office of the Future," *Fortune,* May 18, pp. 44–52.

Uttal, Bro (1983). "The Lab That Ran Away from Xerox," *Fortune,* Sept. 5, pp. 97–102.

Utterback, J. M. (1986). "Innovation and Corporate Strategy," *International Journal of Technology Management* **1**, pp. 119–132.

Walker, R. Charles, and Robert H. Guest (1952). *The Man on the Assembly Line.* Harvard University Press, Cambridge, MA.

Ward, E. Peter (1981). "Planning for Technological Innovation—Developing the Necessary Nerve." *Long Range Planning* **14** (April), pp. 59–71.

Waterman, Jr., R. H. (1987). *The Renewal Factor,* New York, Bantam Books.

Webber, Ross A. (1975). *Management,* Richard D. Irwin, Homewood, IL.

Whisler, Thomas L. (1970). *The Impact of Computers on Organizations,* Praeger, New York.

Whyte, William Foote. (1961). *Men at Work,* Dorsey Press, Homewood, IL.

Woodman, R. W., and J. J. Sherwood (1980). "The Role of Team Development in Organizational Effectiveness: A Critical Review," *Psychology Bulletin* **88** (1).

Woodson, W. E. (1981). *Human Factors Design Handbook,* McGraw-Hill, New York.

Woodson, W. E., and D. W. Conover (1964). *Human Engineering Guide for Equipment Designers,* 2nd ed., University of California Press, Berkeley.

Woodward, Joan, Ed. (1958). *Management and Technology,* Her Majesty's Stationery Office, London.

Woodward, Joan (1965). *Industrial Organization: Theory and Practice.* Oxford University Press, London.

Woodward, Joan (1970). *Industrial Organization: Behavior and Control.* Oxford University Press, London.

FURTHER READING

Abernathy, William J. *The Productivity Dilemma*. Baltimore: Johns Hopkins University Press, 1978.

Abernathy, William J., and J. M. Utterback "Patterns of Industrial Innovation." *Technology Review*, **80**, June–July, pp. 40–47, 1978.

Abernathy, William J., and Kenneth Wayne. "Limits of the Learning Curve." *Harvard Business Review*, September–October, 1974.

Adlard, E. J. "Computer-Integrated Manufacturing—Its Application and Justification." *Engineering Digest*, pp. 16–20, January 1984.

Ahl, David H. "The First Decade of Personal Computing." *Creative Computing*, **10** (11), pp. 30–45, 1984.

Airey, J. "Economic Justification—Counting the Strategic Benefits." *Proceedings of 2nd International Conference on FMS*. London: October 26–28, 1983.

Aldrich, Howard E. "Technology and Organizational Structure: A Reexamination of the Findings of the Aston Group." *Administrative Science Quarterly*, March 1972.

Allision, David, ed. *The R&D Game: Technical Men, Technical Managers and Research Productivity*, Cambridge: MIT Press, 1969.

Anderman, S. D., ed. *Trade Unions and Technological Change*, London: Allen & Unwin, 1967.

Andress, Frank J. "The Learning Curve as a Production Tool." *Harvard Business Review*, January–February 1954.

Andrews, P. P. "Justification Considerations for Robotic Systems." *IBM Robotic Assembly Institute Presentation*, February 1983.

Aronson, Robert. *Jobs, Wages and Changing Technology*. Bulletin 22. Ithaca, NY: Cornell University, School of Industrial and Labor Relations, 1965.

Ashford, Nicholas. *Crisis in the Work Place: Occupational Disease and Injury*. Cambridge, MA: MIT Press, 1976.

Ayres, Robert U. *Technological Forecasting and Long-Range Planning.* New York: McGraw-Hill, 1969.

Backman, Jules, ed. *Labor, Technology, and Productivity.* New York: New York University Press, 1975.

Barbash, Jack. "The Impact of Technology on Labor-Management Relations." In *Adjusting to Technological Change,* Gerald G. Somers et al., eds. New York: Harper & Row, 1963.

"Battle Lines Are Drawn Over Environmental Laws." *Iron Age,* May 12, 1980.

Beattie, C. J., and R. D. Reader *Quantitative Management in R&D.* London: Chapman & Hall, 1971.

Behuniak, J. A. "Economic Analysis of Robot Applications." *SME Technical Paper MS79–777,* 1979.

Bell, James R. "Patent Guidelines for Research Managers." *IEEE Transactions on Engineering Management,* **EM-31** (3), pp. 102–104, 1984.

Benbasat, Izak, and Roger G. Schroeder, "An Experimental Investigation of Some Management Information Systems Variables." *MIS Quarterly,* March 1977.

Ben-David, Joseph. *The Scientist's Role in Society.* Englewood Cliffs, NJ: Prentice-Hall, 1971.

Berg, Norman. "Strategic Planning in Conglomerate Companies. " *Harvard Business Review,* p. 127, May–June 1965.

Bhalla, A. S. *Technology and Employment in Industry: A Case Study Approach.* Geneva: International Labor Organization, 1975.

Bhalla, A. S. "Technology and Employment: Some Conclusions." *International Labour Review,* March–April 1976.

"Biotech Comes of Age." *Business Week.* pp. 84-94, January 23, 1984.

Birnbaum, Philip H. "Strategic Management of Industrial Technology: A Review of the Issues." *IEEE Transactions on Engineering Management,* **EM-31** (4), pp. 186–191, 1984.

Blanchard, Benjamin S. *Engineering Organization and Management.* Englewood Cliffs, NJ: Prentice-Hall, 1976.

Blauner, Robert. "Work Satisfaction and Industrial Trends." In *Labor and Trade Unionism,* Walter Galeman and Seymour Lipset, eds. New York: Wiley, 1960.

Bluestone, Irving. "Emerging Trends in Collective Bargaining." In *Work in America: The Decade Ahead,* Clark Kerr and Jerome M. Rosow, eds. Work in America Institute/Van Nostrand Reinhold Series. New York: Van Nostrand Reinhold, 1979.

Brandt, Steven C. "Strategic Planning in Emerging Companies." New York: Addison-Wesley, 1981.

Braverman, Harry. *Labor and Monopoly Capital.* New York: Monthly Review Press, 1974.

Bright, James R. *Automation and Management.* Boston: Harvard University, Graduate School of Business Administration, Division of Research, 1958.

Bright, James R., and Milton Schoeman. *A Guide to Practical Technological Forecasting.* Englewood Cliffs, NJ: Prentice-Hall, 1973.

Burns, Tom, and G. M. Stalker, *The Management of Innovation.* London: Tavistock Press, 1961.

Carlsson, J., and H. Selg "Swedish Industries' Experience With Robots." *The Industrial Robot,* pp. 88–91, June 1982.

Cassell, Frank H. "Corporate Manpower Planning and Technological Change at the Plant Level." In *Adjustment of Workers to Technological Change at the Plant Level*. Supplement to final report, Organization for European Commercial Development, International Conference, Amsterdam, 1966.

Chadwick-Jones, J. K. *Automation and Behavior*. New York: Wiley, 1969.

Child, John, and Roger Mansfield. "Technology, Size and Organization Structure." *Sociology*, September 1972.

Chinoy, Ely. "Manning the Machines: The Assembly-Line Worker." In *The Human Shape of Work*, Peter Berger, ed. New York: Macmillan, 1964.

Close, D. E. "Productivity & Technology—Top Management's Vital Role." *Manufacturing Engineering*, pp. 101–103, April 1984.

Cole, Robert. "The Day Shift." *Working papers*, July–August, 1979.

Cubertson, Katherine, and Mark Thompson. "An Analysis of Supervisory Training Needs." *Training and Development Journal*, February 1980.

Danzig, Selig M. "How to Bridge the Technology Gap in Manpower Planning." *Management Review*, pp. 18–23, April 1982.

Davies, Celia, et al. "Technology and Other Variables: Some Correct Approaches in Organization Theory." In *Sociology of the Work-Place*, Malcolm Warner, ed. London: Allen & Unwin, 1973.

Day, George S. "A Strategic Perspective on Product Planning." *Journal of Contemporary Business*, pp. 1–34, Spring 1975; reprinted in *Readings in the Management of Innovation*, M. L. Tushmand and W. L. Moore, eds. Marshfield, MA: Pitman, 1982.

Dean, B. V. , and J. L. Goldhar. "Management of Research and Innovation, TIMS Studies in the Management Sciences, Vol. 15." New York: North-Holland, 1980.

Dempsey, P. A. "New Corporate Perspectives in FMS." *Proceedings of 2nd International Conference on FMS*. London, October 26–28, 1983.

Didick, B. J. , ed. *Auto Work and Its Discontents*.Baltimore: Johns Hopkins University Press, 1976.

Didrichsen, Jon. "The Development of Diversified and Conglomerate Firms in the United States 1920–1970." *Business History Review*, **46**, p. 210, Summer 1972.

Drucker, Peter F. "Innovation and Entrepreneurship: Practice and Principles." New York: Harper & Row, 1985.

Drucker, Peter F. "Managing in Turbulent Times." New York: Harper & Row, 1980.

Drucker, Peter F. "Our Entrepreneurial Economy." *Harvard Business Review*, pp. 58–59, January–February 1984.

Dunlop, J. T. , ed. *Automation and Technological Change*. Englewood Cliffs, NJ: Prentice-Hall, 1962.

Eckstein, Otto. "Perspectives on Employment Under Technological Change. " In *Employment Problems of Automation and Advanced Technology*, Jack Stieber, ed. London: Macmillan & Co.; New York: St. Martin's Press, 1966.

Edge, D. O., and J. N. Wolfe, eds. *Meaning and Control*. London: Tavistock Publications, 1973.

Eirma Working Group. "Quality Assurance: R&D and Production/Marketing. *Research Management*, **25**, (5), pp. 25–31, 1982.

Engleberger, J. F. "Robotics in '84." *The Industrial Robot*, September 1979.

Engleberger, J. F. *Robotics in Practice*, pp. 101–110, Amacom, 1980.

Evan, William M. "On the Margin—the Engineering Technician." In *The Human Shape of Work*, Peter Berger, ed., pp. 83–112. New York: Macmillan, 1964.

Fagen, M. D., ed. "A History of Engineering and Science in the Bell System: The Early Years (1875–1925)." Murray Hill, NJ: Bell Telephone Laboratories, 1975.

Fast, Norman D. "Pitfalls of Corporate Venturing." *Research Management,* **24** (2), pp. 21–24, 1981.

Faunce, William A. "Automation and the Division of Labor." *Social Problems,* Fall 1965.

Fenshaw, Peter J., and Douglas Hooper. *The Dynamics of a Changing Technology.* London: Tavistock Publications, 1964.

Fleck, J. "The Adoption of Robots." *Proceedings of Robots 7,* pp. 1.41–1.51, Chicago, Illinois, April 17–21, 1983.

Ford, David, and Chris Ryan. "Taking Technology to Market." *Harvard Business Review,* pp. 18–23, March–April 1981.

Form, William H. "Auto Workers and Their Machines: A Study of Work, Factory and Job Satisfaction in Four Countries." *Social Forces,* September 1973.

Form, William H. *Blue-Collar Stratification: Autoworkers in Four Countries.* Princeton: Princeton University Press, 1976.

Forrester, Jay W. "Innovation and the Economic Long Wave." *Management Review,* pp. 16–24, June 1979.

Foster, Richard N. "A Call for Vision in Managing Technology." *Business Week,* pp. 24–33, May 24, 1982.

Foster, Richard N. "Organize for Technology Transfer." *Harvard Business Review,* November–December 1971.

Fraker, Susan. "High-Speed Management for the High-Tech Age." *Fortune,* pp. 62–68, March 5, 1984.

Fullan, Michael. "Industrial Technology and Worker Integration in the Organization." *American Sociological Review,* December 1970.

Gee, Edwin A. *Managing Innovation.* New York: Wiley, 1976.

General Electric Motor Company. *Automation in General Electric: The Human Side of the Story.* General Electric Motor Company, 1964.

Gerstenfeld, A., ed. *Technological Innovation: Government/Industry Cooperation,* New York: Wiley, 1977.

Ginzberg, Eli, ed. *Technology and Social Change.* New York: Columbia University Press, 1964.

Gluck, Frederick W., Stephen P. Kaufman, and Stephen A. Walleck. "Strategic Management for Competitive Advantage." *Harvard Business Review,* pp. 154–161, July–August 1980.

Gold, Bela. "CAM Sets New Rules for Production." *Harvard Business Review,* pp. 88–94, November–December 1982.

Gold, Bela, ed. *The Productivity Dilemma: Roadblock to Innovation in the Automated Industry.* Baltimore: Johns Hopkins University Press, 1978.

Gold, Bela, ed. *Technological Change: Economics, Management and Environment.* New York: Pergamon Press, 1975.

Gomberg, William. "The Work Rules and Work Practices Problem." *Labor Law Journal,* July 1961.

Gordon, M. S. "The Comparative Experience with Retraining Programs in the U.S. and Europe." In *Employment Problems of Automation and Advanced Technology,* Jack Stieber, ed. London: Macmillan & Co.; New York: St. Martin's Press, 1966.

Gottlieb, Daniel. "Technology Training Surges." *High Technology,* pp. 70–73, October 1983.

Govsievich, R. E. "Determining the Economic Effectiveness of Industrial Robots." *Machines & Tooling,* **49** (8), pp. 11–13, 1978.

Gray, Steven B. "The Early Days of Personal Computers." *Creative Computing,* **10** (11), pp. 6–14, 1984.

Green, Alice M. "MIS Responds to Today's Company Needs." *Iron Age,* March 17, 1980.

Grimes, A. J., and S. M. Klein. "The Technological Imperative: The Relative Impact of Task Unit Model Technology and Hierarchy on Structure." *Academy of Management Journal,* December 1973.

Guest, Robert. *Organizational Change: The Effect of Successful Leadership.* Homewood, IL: Dorsey Press, 1962.

Gyllenhammar, Peter G. "How Volvo Adapts Work to People." *Harvard Business Review,* July–August 1977.

Haber, William, et al. *The Impact of Technological Change.* Kalamazoo, MI: W. E. Upjohn Institute, 1963.

Haeffner, Erik. "Critical Activities of the Innovation Process." In *Current Innovation,* B. A. Vedin, ed., pp. 129–144. Stockholm: Almquist & Wiksell.

Hartley, John. "The Japanese Scene: Developments in Japan." *The Industrial Robot,* March 1980.

Harvey, Edward. "Technology and the Structure of Organization." *American Sociological Review,* April 1968.

"Has OSHA Become Part of National Consciousness?" *Iron Age,* April 28, 1980.

Hayes, Robert, and David Garvin. "Managing As If Innovation Mattered." *Harvard Business Review,* pp. 70–79, May–June, 1982.

Hayes, Robert H., and Steven C. Wheelwright. "The Dynamics of Process-Product Life Cycles." *Harvard Business Review,* p. 127, March–April, 1979.

Henwood, Felicity, and Graham Thomas. *Science, Technology and Innovation: A Research Bibliography.* New York: St. Martin's Press, 1983.

Hicks, Wayland R. "A New Approach to Product Development." *High Technology,* pp. 11–12, October 1984.

Hickson, David J., et al. "Operations Technology and Organization Structure: An Empirical Reappraisal." *Administrative Science Quarterly,* March 1974.

Hildebrandt, George H. "Some Alternative Views of the Unemployment Problem in the U.S." In *Employment Problems of Automation and Advanced Technology,* Jack Steiber, ed. London: Macmillan & Co.; New York: St. Martin's Press, 1966.

Hill, Christopher T., and James M. Utterback eds. "Technological Innovation for a Dynamic Economy." Elmsford, NY: Pergamon Press, 1979.

Hirschmann, Winfred B. "Profit from the Learning Curve." *Harvard Business Review,* January–February 1964.

Hower, R., and Charles D. Orth, Charles D. *Managers and Scientists: Some Human Problems in Industrial Research Organizations.* Boston: Harvard University, Graduate

School of Business Administration, Division of Research, 1963.

Inagaki, Souji. "Standardization of Industrial Robots Question." *The Industrial Robot,* March 1980.

Industrial Relations Research Association. *Collective Bargaining and Productivity.* Madison, WI: University of Wisconsin, 1975.

International Conference on Automation, Full Employment and a Balanced Economy. Conference papers. New York: American Foundation on Automation and Employment, 1967.

International Labor Organization. *Manpower Adjustment Programs: II,* Section on United States. Bulletin 6. Geneva, 1969.

Irving, Robert R. "Technical Societies Rise to the New Challenge." *Iron Age,* November 13, 1978.

Irving, Robert R. "Why America's Technology Is in Trouble and What Can Be Done About It." *Iron Age,* July 31, 1978.

Jaffe, A. J., and Joseph Foomkin. *Technology and Jobs.* New York: Praeger, 1968.

Jehring, J. J., ed. *Productivity and Automation.* National Council for Social Studies. Washington, D.C.: U.S. Government Printing Office, 1966.

Jelinek, Mariann. *Institutionalizing Innovation: A Study of Organizational Learning Systems.* New York: Praeger, 1979.

Johnson, Nils O., and D. B. Davidson. "Realigning and R&D Organization From R-Intensive to D-Intensive: A Case Example." *IEEE Trans. on Engineering Management,* **EM-29** (1), February 1982.

Judson, Arnold S. "New Strategies to Improve Productivity. *Technology Review,* July-August, 1976.

Judson, Horace Freeland. it The Eighth Day of Creation. A Study of Organizational Learning Systems.New York: Simon and Schuster, 1979.

Kalachek, Edward D. *Labor Markets and Unemployment.* Belmont, CA: Wadworth, 1973.

Kantrow, Alan M. "The Strategy-Technology Connection." *Harvard Business Review,* pp. 6–21, July-August 1980.

Karger, Theodore. "Defining a Leadership Mandate." *Management Review,* pp. 14–17, April 1982.

Kaye, Beverly. "Career Development Puts Training In Its Place." *Personnel Journal,* February 1983.

Kennedy, Thomas. *Automation Funds and Displaced Workers.* Boston: Harvard University, Graduate School of Business Administration, Division of Research, 1962.

Khandwallah, Pralip N. "Mass Output Orientation of Operations Technology and Organizational Structure." *Administrative Science Quarterly,* March 1974.

Killingsworth, Charles C. "Cooperative Approaches to Problems of Technological Change." In *Adjusting to Technological Change,* Gerald G. Somers, et al., eds. New York: Harper & Row, 1963.

Killingsworth, Charles C. "Structural Unemployment in the U.S." In *Employment Problems of Automation and Advanced Technology,* Jack Stieber, ed. London: Macmillan & Co.; New York: St. Martin's Press, 1966.

Killingsworth, Charles C. "Structural Unemployment Without Quotation Marks." *Monthly Labor Review,* June 1979.

Kotter, John P. and L. A. Schlesinger. "Choosing Strategies for Change." *Harvard Business Review,* March-April 1979.

Kreps, Juanita, and Ralph Laws. *Automation and the Older Worker: An Annotated Bibliography*. New York: National Council on Aging, 1963.

Lanahan, John R. "Information Systems at Inland Steel. *MIS Quarterly*, June 1978.

Larsen, R. J. "Group Technology Pierces the Manufacturing Horizon." *Iron Age*, November 13, 1978.

Lawrence, Paul. "The Harvard Organization and Environmental Research Program." Paper presented at the Organization Behavior Conference, University of Pennsylvania, 1980.

Lawrence, Paul, and John Seiler. *Organizational Behavior and Administration*, rev. ed. Homewood, IL: Dorsey Press, 1965.

Levien, Roger E., et al. "The Emerging Technology—Instructional Uses of the Computer in Hyphen Education." New York: McGraw-Hill, 1972.

Levine, Susan J., and Michael S. Yalowitz. "Managing Technology: The Key to Successful Business Growth." *Management Review*, pp. 43–38, September 1983.

Likert, Rensis. *New Patterns of Management*. New York: McGraw-Hill, 1961.

Link, Albert N. "Impact of Federal Research and Development Spending on Productivity." *IEEE Trans. on Engineering Management*, **EM-29** (4), pp. 166-168, November 1982.

Litecky, C. R. "Intangibles in Cost-Benefit Analysis." *Journal of Systems Management*, pp. 15–17, February 1981.

Lohr, Steve. "The New Industrial Robots Are Punching Clocks Faster." *The New York Times*, April 27, 1980.

"Longbridge Robots Will March Over the Transport Union." *The Economist*, April 19, 1980.

Lorsch, Jay W. *Organization and Environment: Managing Differentiation and Integration*. Boston: Harvard University, Graduate School of Business Administration, Division of Research, 1967.

Lorsch, Jay W. *Studies in Organizational Design*. Homewood, IL: Dorsey Press, 1970.

Lorsch, Jay W., and Paul R. Lawrence. "Environmental Factors and Organizational Integration." In *Organizational Planning*, Jay Lorsch and Paul Lawrence, eds. Homewood, IL: Dorsey Press, 1972.

Lynch, Beverly P. "An Empirical Assessment of Perrow's Technology Construct." *Administrative Science Quarterly*, September 1974.

McCarthy, James E. *Trade Adjustment Assistance: A Case Study of the Shoe Industry in Massachusetts*. RR58. Boston: Federal Reserve Bank, 1975.

McDonald, J. L., and W. F. Hastings. "Selecting and Justifying CAD/CAM." *Assembly Engineering*, pp. 24–27, April 1983.

McDonough, F. Edward III, and Raymond M. Kinnunen. "Management Control of New Product Development Projects." *IEEE Transactions on Engineering Management*, **EM-31** (1), pp. 18–21, 1984.

Mahoney, Thomas A., and Peter J. Frost. "The Role of Technology in Models of Organizational Effectiveness." *Organizational Behavior and Human Performance*, February 1974.

Maidique, Modesto A. "Entrepreneurs, Champions, and Technological Innovation." *Sloan Management Review*. 21 (2), pp. 59–76, 1980; reprinted in *Readings in the Management of Innovation*, M. L. Tushman and W. L. Moore, eds. Marshfield, MA: Pitman, 1982.

Mansfield, Edwin. *The Economics of Technological Change*. New York: W. W. Norton, 1968.

Mansfield, Edwin. "How Economists See R&D." *Harvard Business Review,* pp. 98–106, November–December 1981.

Mansfield, E., et al. *The Production and Application of New Industrial Technology.* New York: Norton & Co., 1977.

Markham, Charles, ed. *Jobs, Men and Machines: Problems of Automation.* New York: Praeger, 1964.

Meadows, Edward. "A Close-Up Look at the Productivity Lag." *Fortune,* December 4, 1978.

Meissner, Martin. *Technology and the Worker.* San Francisco: Chandler, 1969.

Merton, Robert K. *Social Theory and Social Structure.* New York: The Free Press, 1957.

Meyer, R. J. "A Cookbook Approach to Robotics and Automation Justification." *Proceedings of Robots 6.* Detroit, Michigan, March 2–4, 1982.

Michaels, L. T., W. T. Muir, and R. G. Eiler. "Improving Technology Cost-Benefit Analysis." *Material Handling Engineering,* pp. 49–54, February 1984.

Michaels, L. T., et al. "The Relationship Between Technology and Cost Management." *Material Handling Engineering,* pp. 58–64, January 1984.

Miller, Bruce H. "Providing Assistance to Displaced Workers." *Monthly Labor Review,* May 1979.

Miller, Jeffrey G., and Linda G. Sprague. "Behind the Growth in Materials Requirements Planning." *Harvard Business Review,* September–October 1975.

Miller, Marc. "How to Manage the Work Ethic in the Automated Workplace." *Management Review,* pp. 8–12, September 1983.

Miller, R. K. "The Bottom Line—Justifying a Robot Installation." *Robotics World,* pp. 32–35, April 1983.

Mueller, Eva, et al. *Technological Advance in an Expanding Economy.* Ann Arbor, MI: University of Michigan Press, 1969.

Muir, W. T. "Cost-Benefit Analysis and Cost-Benefit Tracking." *Proceedings of AUTOFACT 5,* pp. 8.34–8.47, Detroit, Michigan, November 14–17, 1983.

Murphy, Sheldon R. "Five Ways to Improve R&D Efficiency." *Research Management,* (1), pp. 8–0, 1981.

Myers, John G., et al. *Energy Consumption in Manufacturing.* New York: The Conference Board, 1974.

Naidish, N. L. "Justifying Assembly Automation, Part 1." *Assembly Engineering,* pp. 46–49, April 1982.

Nadish, N. L. "Justifying Assembly Automation, Part 2." *Assembly Engineering,* pp. 48–49, May 1982.

National Academy of Engineering and National Research Council. *The Competitive Status of the U.S. Pharmaceutical Industry, 1983.* Washington, D.C.: National Academy Press, 1983.

National Center for Productivity and Quality of Working Life. *Productivity in the Changing World of the 1980's,* Final Report. Washington, D.C.: U.S. Government Printing Office, 1978.

National Center for Productivity and Quality of Working Life. *Productivity and Job Security: Attrition Benefits and Problems.* Washington, D.C.: U.S. Government Printing Office, 1977.

National Center for Productivity and Quality of Working Life. *Productivity and Job Security: Retraining to Adapt to Technological Change.* Washington, D.C.: U.S. Government Printing Office, 1977.

National Center for Productivity and Quality of Working Life. *Recent Initiatives in Labor-Management Cooperation,* **2,** Washington, D.C.: U.S. Government Printing Office, 1978.

National Center for Productivity and Quality of Working Life. *A Summary of the Future of Productivity,* **1,** Washington, D.C.: U.S. Government Printing Office, 1978.

National Commission on Technology, Automation and Economic Progress Report. Washington, D.C.: U.S. Government Printing Office, 1966.

National Science Board, National Science Foundation. Science Indicators 1982. Washington, D.C.: Superintendent of Documents, U.S. Government Printing Office, 1983.

Nelson, Richard R., et al. *Technology, Economic Growth and Public Policy.* Washington, D.C.: The Brookings Institution, 1967.

Noble, David. *America by Design.* New York: Oxford University Press, 1977.

Noble, David. *Social Choice in Machine Design: The Case of Automatically Controlled Machine Tools.* Cambridge, MA:
Massachusetts Institute of Technology, School of Humanities and Social Sciences.

Orth, Charles D., et al., eds. *Administering Research and Development: The Behavior of Scientists and Engineers in Organizations.* Homewood, IL: Irwin, 1964.

Palm, Goran. *The Flight from Work.* New York: Columbia University Press, 1977.

Paolillo, Joseph G. P. "Technological Gatekeepers: A Managerial Perspective." *IEEE Trans. on Engineering Management,* **EM-29** (4), November 1982.

Perrow, Charles. "A Framework for the Comparative Analysis of Organizations." *American Sociological Review,* April 1967.

Petit, Thomas A. "A Behavioral Theory of Management. " *Journal of the Academy of Management,* December 1967.

Potter, R. D. "Analyze Indirect Savings in Justifying Robots." *Industrial Engineering,* pp. 28, 29, 94, November 1983.

Powell, R. E. "Justification and Financial Analysis for CAD." *Proceedings of the 17th Design Automation Conference,* pp. 564-471, Minneapolis, Minnesota, June 23–25, 1980.

Pyke, Donald L. "Technological Forecasting: A Framework for Consideration." *Futures,* p. 327, December 1970.

Quinn, James Brian. *Strategies for Change: Logical Incrementalism.* Homewood, IL: Dow Jones-Irwin, 1980.

Quinn, James Brian. "Technological Innovation, Entrepreneurship, and Strategy." *Sloan Management Review,* **20** (3), 19–30, 1979; reprinted in *Readings in the Management of Innovation,* Tushman, M. L., and Moore, W. L., eds. Marshfield, MA: Pitman, 1982.

Reeves, Tom Kynaston, and Woodward, Joan. "The Study of Managerial Control." In *Industrial Organization: Behavior and Control,* Joan Woodward, ed. London: Oxford University Press, 1970.

Rezler, John. *Automation and Industrial Labor.* New York: Random House, 1969.

Robinson, Arthur L. "One Billion Transistors on a Chip?" *Science,* **223** pp. 73–84, January 20, 1984.

"Robots Change the Rules." *The Economist*, April 19, 1980.

"Robots Make Economic and Social Sense." *Atlanta Economic Review*, July–August 1977.

Rodgers, R. C. "How to Join the Robot Revolution—Part I." *Foundry M&T*, pp. 24–26, March 1983.

Roethlisberger, F. "The Foreman: Master and Victim of Double-Talk." *Harvard Business Review*, March–June 1945.

Rogers, P. F. "The Economics of Robotic Arc Welding Workcells." *Robotics Today*, pp. 46–48, June 1984.

Roman, Daniel D. *Science, Technology, and Innovation: A Systems Approach*. Columbus, Ohio: Grid Publishing, 1980.

Rosenberg, Jerry M. *Automation, Manpower and Education*. New York: Random House, 1966.

Rosenberg, Nathan. *Perspectives on Technology*. New York: Columbia University Press, 1976.

Rosenbloom, Richard S. *Technological Innovation in Firms and Industries: An Assessment of the State of the Art, in Technological Innovation*. Kelly, P. and Kranzberg, M., eds. San Francisco: San Francisco Press.

Rosenbloom, Richard S., and William J. Abernathy. "The Climate for Innovation in Industry." *Research Policy*, **11** (4), pp. 218–225, 1982.

Rothermel, Terry W. "Forecasting Resurrected." *Harvard Business Review*, pp. 139–147, March–April, 1982.

Roussel, Philip A. "Cutting Down the Guesswork in R&D." *Harvard Business Review*, pp. 139–147, March–April, 1982.

Ruggles, Rudy L. "How to Integrate R&D and Corporate Goals." *Management Review*. September 1982.

Sappho-Report on Project Sappho. "Success and Failure in Industrial Innovation." Center for the Study of Industrial Innovation, 162 Regent St, London, W1R 6DD 2/72, 1972.

Sayles, Leonard R. *Behavior of Industrial Work Groups: Prediction and Control*. New York: Wiley, 1963.

Scanlan, Burt, and J. Benard Keys. *Management and Organizational Behavior*. John Wiley and Sons, 1979.

Schmidtt, Roland W. "Successful Corporate R&D." *Harvard Business Review*, pp. 124–129, May–June 1985.

Schon, D. A. "Champions for Radical New Inventions." *Harvard Business Review*, March–April, 1985.

Schrank, Robert. *Ten Thousand Working Days*. Cambridge, MA: MIT Press, 1978.

Schultz, George P., and Arnold R. Wever. *Placement and Retraining Experience with Workers Displaced by Shutdown of the Main Armour Plant in Fort Worth, Texas*. Report to the Automation Fund Committee. Fort Worth, TX: Armour and Company, 1963.

Scott, Ellis L., and Roger W. Bolz. *Automation Management: The Social Perspective*. Athens, GA: Center for the Study of Automation and Society, 1970.

Scott, Ellis L., and Roger W. Bolz. *Automation and Society*. Athens, GA: Center for the Study of Automation and Society, 1969.

Seeman, Melvin. "On the Meaning of Alienation." In *Automation, Alienation and Anomie*, Marscon, Simon, ed. New York: Harper & Row, 1977.

Sepehri, M. "Cost Justification Before Factory Automation." *P&IM Review and APICS News*, pp. 40–48, April 1984.

Shapiro, Reuven, and Shlomo Gloverson. "An Incentive Plan for R&D Workers." *Research Management*, pp. 17–20, September–October 1983.

Shepard, Jon. *Automation and Alienation*. Cambridge, MA: MIT Press, 1970.

Shepard, Jon. "Functional Specialization, Alienation and Job Specialization." *Industrial and Labor Relations Review*, January 1970.

Sheriff, Don R. , et al., eds. *The Social and Technological Revolution of the Sixties*. Iowa City: University of Iowa, College of Business Administration, Bureau of Labor and Management, 1965.

Smith, George David, and Lawrence E. Steadman. "Present Value of Corporate History." *Harvard Business Review*, p. 164, November–December 1981.

Smith, J. J. , et al. "Lessons From Ten Case Studies in Innovation." *Research Management*, **22** (5), pp. 23– 27, 1984.

Smith, Lee. "The Lures and Limits of Innovation." *Fortune*, pp. 84–94, October 20, 1980.

Society of Manufacturing Engineers. *The Manufacturing Engineer: Past, Present, and Future*, Columbus, OH: Society of Manufacturing Engineers, 1979.

Sorcher, Melvin. "Motivation on the Assembly Line." *Personnel Administration*, May–June 1969.

Staples, G. R. "FMS—Convincing the Board." *Proceedings of 2nd International Conference on FMS*. London, UK, October 26–28, 1983.

"Steel Labor Talks Are Vital to Survival Plan." *Iron Age*, February 25, 1980.

Taylor, J. W., and N. J. Dean. "Managing to Manage the Computer." *Harvard Business Review*, September–October 1966.

Tepsic, R. M. "How to Justify Your FMS." *Manufacturing Engineering*, pp. 50–52, September 1983.

Thurley, Keith A. "Computers and Supervisors." In *Sociology of the Work-Place*, edited by Malcolm Warner. London: Allen & Unwin, 1973.

Tornatzky, Louis G., and K. J. Klein. "Innovation Characteristics and Innovation Adoption-Implementation: A Meta Analysis of Finding." *IEEE Trans. on Engineering Management*. **EM-29** (1), pp. 28–45, February 1982.

Trist, E. L., and K. W. Bamforth. "The Long-Wall Method of Coal Getting." In *Industrial Man*, Burns, Tom, ed. New York: Penguin Books, 1969.

Turner, Arthur N., and Paul R. Lawrence. *Industrial Jobs and the Worker: An Investigation of Response to Task Attributes*. Boston: Harvard University, Graduate School of Business Administration, 1965.

Tushman, M. L., and R. Katz. "External Communication and Project. Performance: An Investigation Into the Role of Gatekeepers." *Management Science*, **26** (11), pp. 1071–1085, 1985.

Tushman, Michael L., and W . L. Moore, eds. *Readings in the Management of Innovation*. Marshfield, MA: Pitman, 1982.

Udy, Stanley, H., Jr. "The Comparative Analysis of Organization." In *Handbook of Organization*, ed. Chicago: Rand McNally, 1965.

Urisko, James A. "Productivity in Grain Mill Products." *Monthly Labor Review*, April 1977.

U.S. Comptroller General, General Accounting Office. *Manufacturing Technology: A Changing Challenge to Improved Productivity.* Washington, D.C.: U.S. Government Printing Office, 1976.

U.S. Department of Labor, Bureau of Labor Statistics. *Improving Productivity.*Bulletin 1715. Washington, D.C.: U.S. Government Printing Office, 1972.

U.S. Department of Labor, Bureau of Labor Statistics. *Job Redesign for Older Workers: Ten Case Studies.* Bulletin 1523. Washington, D.C. : U.S. Government Printing Office, 1966.

U.S. Department of Labor, Bureau of Labor Statistics. *Layoffs, Recalls and Work-Sharing Procedures.* Bulletin 1423-13. Washington, D.C.: U.S. Government Printing Office, 1972.

U.S. Department of Labor, Bureau of Labor Statistics. *Manpower Implications of Automation.* Washington, D.C.: U.S. Government Printing Office, 1964.

U.S. Department of Labor, Bureau of Labor Statistics. *Productivity and Job Security: Attrition Benefits and Problems.* Washington, D.C: U.S. Government Printing Office, 1977.

Utterback, James. "Management of Technology", in *Studies in Operation Management.* Arnoldo C. Hax, eds. Amsterdam:
North-Holland Publishing, pp. 137–160, 1978.

VanBlois, J. P. "Economic Models: The Future of Robotic Justification." *Proceedings of Robot 7,* pp. 4.24–4.31, Chicago, Illinois, April 17–21, 1983.

VanBlois, J. P. "Robotic Justification Considerations." *Proceedings of Robots 6.* Detroit, Michigan, March 2–4, 1982.

VanBlois, J. P., and Andrews, P. P. "Robotic Justification: The Domino Effect." *Production Engineering,* pp. 52–54, April 1983.

Walker, Charles R. "Basic Human Problems in Mass-Production Technologies: Pacing, Pressure and Repetitiveness." In *Technology, Industry and Man: The Age of Acceleration,* Walker, Charles R., ed. New York: McGraw-Hill, 1968.

Walker, Charles R. *Toward the Automatic Factory.* New Haven: Yale University Press, 1957.

Walton, Richard. *A Developmental Theory of High- Commitment Work Systems.* Boston: Harvard University, Graduate School of Business Administration, Division of Research, 1979.

Walton, Richard. "How to Counter Alienation in the Plant." *Harvard Business Review,* November–December 1972.

Washington, State of, Employment Security Department. *Technology Changes in the Plywood Occupations.* Seattle, 1968.

Wever, Arnold R. "The Interplant Transfer of Displaced Employees." In *Adjusting to Technological Change,* Gerald G. Somers, et al., eds. New York: Harper & Row, 1963.

Wheelwright, Steven C., and Darral G. Clarke. "Corporate Forecasting Promise and Reality." *Harvard Business Review,* pp. 100–107, November–December 1976.

Wheelwright, Steven C., and Robert H. Hayes. *Restoring Our Competitive Edge: Competing Through Manufacturing.* New York: Wiley, 1984.

White, George R., and Margaret B. W. Graham. "How to Spot a Technological Winner." *Harvard Business Review,* pp. 146–152, March–April 1978.

Wild, Ray. *Work Organization: A Study of Manual Work and Mass Production.* New York: Wiley, 1975.

"Will Alcoa Process 'Deck' the Hall?" *Iron Age,* January 1973.

Wolff, Michael. "The Why, When and How of Directed Basic Research." *Research Management,* **24** (3), pp. 29–31, 1981.

Wolff, Michael F., Mervin J. Kelly. "Manager and Motivator." *IEEE Spectrum,* pp. 71–75, December 1983.

"World's Fastest Rolling Mill." *Light Metal Age,* April 1972.

York, James D. "Folding Paperboard Box Industry Shows Slow Rise in Productivity." *Monthly Labor Review,* March 1980.

York, James, and Brand, Horst. "Productivity and Technology in the Electric Motor Industry." *Monthly Labor Review,* August 1978.

Young, Edwin. "The Armour Experience: A Case Study in Plant Shut-Down." In *Adjusting to Technological Change,* Somers, Gerald G., et al., eds. New York: Harper & Row, 1963.

Zachary, William B., and Robert M. Krone. "Managing Creative Individuals in High-Technology Research Projects." *IEEE Transactions on Engineering Management,* **EM-31** (1), pp. 37–40, 1984.

Zager, Robert. "The Problem of Job Obsolescence: Working It Out at River Works." *Monthly Labor Review,* July 1978.

Zalewski, C. "The Influence of Automation on Management." In *Employment Problems of Automation and Advanced Technology,* Jack Stieber, ed. London: Macmillan & Co.; New York: St. Martin's Press, 1966.

Zaltman, G. R., Duncan and J. Holbek. *Innovation and Organizations.* New York: Wiley, 1973.

Zwerman, William L. *New Perspectives on Organization Theory.* Westport, CT: Greenwood Press, 1970.

Appendix A
COMPUTER-AIDED MANUFACTURING ERGONOMIC CHECKLIST (CAMEC)

CAMEC is based on two case studies in a computer-aided manufacturing environment. The checklist should be used only as a guide; several other system design requirements that are not specified in CAMEC may be needed. Each particular type of computer-aided system might also require unique sets of features, variables, and parameters that must be taken into account during system design and testing phases. The design rules presented in Section 3.2 must also be used to achieve total system optimization.

_____	Computer speed	_____	Levers/switches
_____	Communication channels	_____	Indicators (visual)
_____	System feedback	_____	System labels
_____	Volume/to GGK switches	_____	System light and pictorial
_____	System variable controls	_____	Printed materials and signals
_____	Task complexity	_____	System relationships
_____	Motions/movements	_____	Indicators (auditory)
_____	Arm rests/gimbal	_____	System pure tones
_____	Pressure controls	_____	System speech pattern
_____	Noise controls	_____	Static proprioceptive
_____	Light controls	_____	Dynamic proprioceptive
_____	Visual display terminals	_____	Systems vibration and intensity
_____	Keyboard height	_____	Systems movement and texture
_____	Adjustable terminal base	_____	Human-computer response time
_____	Optional tilt devices	_____	Delays signals for human interaction
_____	Platform height controls		

APPENDIX A: COMPUTER-AIDED MANUFACTURING ERGONOMIC CHECKLIST

- _____ Heat sensors
- _____ Light sensors
- _____ Wave/static sensors
- _____ Reference stand for materials
- _____ Programming flexibility
- _____ System cables
- _____ System electrical outlets
- _____ System ventilation
- _____ Entry command controls
- _____ System variables/expressions
- _____ Cross-batch menu protection
- _____ Operator get/reach motion
- _____ Error recovery tolerances
- _____ Supporting tools dimensions
- _____ Tool boundary display
- _____ Tool movement
- _____ Tool positioning
- _____ Dimensions of work envelope
- _____ Other dimensions
- _____ Workplace design
- _____ Task design
- _____ Safety devices
- _____ Numerical controls
- _____ Work envelope posture
- _____ Sitting chairs/foot rest
- _____ Back rests/seat pans
- _____ Task content/elements
- _____ System reaction time
- _____ System capability loading
- _____ Environment constraints
- _____ Wave logistics controls
- _____ Malfunction profiles
- _____ Operating consoles
- _____ Other consoles
- _____ Defect control signal intervention
- _____ User-system interface
- _____ System associativity
- _____ Controls for editors/debug
- _____ Error recovery made
- _____ Interruption devices
- _____ Load and unload devices
- _____ Test points
- _____ Access points
- _____ Vibration attenuation
- _____ Radiation intensity
- _____ Simulators
- _____ Push buttons' effectiveness
- _____ Devices' effectiveness
- _____ Signals and direction
- _____ Coded instructions
- _____ Digital controls
- _____ System calibration
- _____ Display angle
- _____ Command modifiers
- _____ Insertion speed
- _____ System snap controls
- _____ Sensors for heat control
- _____ Sensors for electrical controls
- _____ Task decision latitude
- _____ Task structure difficulty
- _____ Task structure coordination
- _____ Task complexity factor
- _____ Task fatigue controls
- _____ Task load and unload controls
- _____ Input and output balancing devices and sensors

Appendix B
COMPARATIVE JUDGMENT INSTRUMENT (CJI) FOR ASSESSMENT OF JOB SATISFACTION AND PSYCHOLOGICAL STRESS

TASK PREFERENCE ANALYSIS SURVEY

Instructions

1. Please do not sign your name; the survey is anonymous.
2. Answer the questions below based on your own overall experience in performing the following two tasks:

   Task M = Assembly Cards Manually

   Task C = Assembly Cards using the (Computer)

Example

WHICH TASK WOULD YOU RECOMMEND TO A FRIEND?

( ) Task M (Manual)   (X) Task C (Computer)

On the following scale of 1–5 rate by how much you recommend the task selected to a friend by writing M (Manual) and C (Computer) in the appropriate spaces.

M (Manual)   C (Computer)

1	2	3	4	5
Would never recommend		Might recommend		Would recommend highly

APPENDIX B: COMPARATIVE JUDGEMENT INSTRUMENT(CJI)

QUESTION 1: WHICH TASK IS MORE INTERESTING?

( ) Task M (Manual)   ( ) Task C (Computer)

On the following scale rate each task to show how interesting it is by writing M and C in the appropriate spaces.

1	2	3	4	5
Very uninteresting		Slightly interesting		Extremely interesting

QUESTION 2: WHICH TASK IS MORE DIFFICULT TO PERFORM?

( ) Task M (Manual)   ( ) Task C (Computer)

Rate each task to show how difficult it is by writing M and C in the appropriate spaces below.

1	2	3	4	5
Very easy	Somewhat easy	Slightly difficult	Somewhat difficult	Very difficult

QUESTION 3: WHICH TASK IS MORE FATIGUING?

( ) Task M (Manual)   ( ) Task C (Computer)

Rate each task to show how fatiguing it is by writing M and C in the appropriate spaces below.

1	2	3	4	5
Hardly fatiguing		Slightly fatiguing		Very fatiguing

QUESTION 4: WHICH TASK DO YOU LIKE THE MOST?

( ) Task M (Manual)   ( ) Task C (Computer)

Rate each task to show how much you like it by writing M and C in the appropriate spaces below.

1	2	3	4	5
Do not like the task at all	Dislike somewhat	Neither like nor dislike very much	Like somewhat	Like the task very much

QUESTION 5: WHICH TASK REQUIRES MORE PHYSICAL EFFORT?

( ) Task M (Manual)   ( ) Task C (Computer)

Rate each task to show how much physical effort it requires by writing M and C in the appropriate spaces below.

1	2	3	4	5
Very little physical effort is required		Moderate physical effort required		High physical effort required

QUESTION 6: WHICH TASK REQUIRES MORE HAND COORDINATION?

( ) Task M (Manual)   ( ) Task C (Computer)

Rate each task to show how much hand coordination it requires by writing M and C in the appropriate spaces below.

1	2	3	4	5
Requires very little hand		Requires some hand coordination		Requires high hand coordination

QUESTION 7: IN WHICH TASK DO YOU HAVE MORE CONTROL OVER WHAT YOU DO?

( ) Task M (Manual)   ( ) Task C (Computer)

Rate each task to show how much control you have in doing it by writing M and C in the appropriate spaces below.

1	2	3	4	5
Very little control	Little control	Moderate control	Fairly good control	High control

QUESTION 8: WHICH TASK REQUIRES MORE CONCENTRATION?

( ) Task M (Manual)   ( ) Task C (Computer)

Rate each task to show how much concentration it requires by writing M and C in the appropriate spaces below.

1	2	3	4	5
Low concentration		Moderate concentration		High concentration

QUESTION 9: WHICH TASK IS MORE BORING?

( ) Task M (Manual)   ( ) Task C (Computer)

Rate each task to show how boring it is by writing M and C in the appropriate spaces below.

1	2	3	4	5
Not boring at all		Somewhat boring		Very boring

QUESTION 10: WHICH TASK IS MORE STRESSFUL?

( ) Task M (Manual)   ( ) Task C (Computer)

Rate each task to show how stressful it is by writing M and C in the appropriate spaces below.

1	2	3	4	5
Not stressful at all	A little stressful	Moderately stressful	Fairly stressful	Very stressful

QUESTION 11: WHICH TASK IS MORE CONFUSING?

( ) Task M (Manual)   ( ) Task C (Computer)

Rate each task to show how confusing it is by writing M and C in the appropriate spaces below.

1	2	3	4	5
Not confusing at all		Somewhat confusing		Very confusing

QUESTION 12: WHICH TASK IS EASIER TO LEARN?

( ) Task M (Manual)   ( ) Task C (Computer)

Rate each task to show how easy it is to learn by writing M and C in the appropriate spaces below.

1	2	3	4	5
Quite difficult to to learn	Somewhat difficult to learn	Slightly difficult to learn	Fairly easy to learn	Very easy to learn

QUESTION 13: WHICH TASK ALLOWS YOU TO MAKE MORE DECISIONS ON YOUR OWN?

( ) Task M (Manual)   ( ) Task C (Computer)

Rate each task to show the extent to which it allows you to make decisions on your own by writing M and C in the appropriate spaces below.

1	2	3	4	5
Allows very few decisions	Allows some decisions	Allows quite a number of decisions	Allows many decisions	Allows a great number of decisions

QUESTION 14: WHICH TASK REQUIRES WORKING WITH THE GREATEST SPEED?

( ) Task M (Manual)   ( ) Task C (Computer)

Rate each task to show the rate of speed required by writing M and C in the appropriate spaces below.

1	2	3	4	5
Low speed		Moderate speed		High speed

QUESTION 15: WHICH TASK FORCES YOU MOST TO RELY ON OTHER PEOPLE (CO-WORKERS, STAFF, MAINTENANCE, DEPT. TECH., ETC.)?

( ) Task M (Manual)   ( ) Task C (Computer)

Rate each task to show how much it forces you to rely on other people by writing M and C in the appropriate spaces below.

1	2	3	4	5
Very little reliance on other people		Moderate reliance on other people		High reliance on other people

INDEX

A

Adoption theory, 212
Anthropometric dimensions, 46
Apple Computer, 146
Attribute list, 156
Automated assembly technologies, 10
Automatic identification technologies, 155
Automation strategy, 247
Auxiliary equipment, 221
Average person, 46

B

Body dimensions, 48–51

C

Calibration, 57
CAMEC, 45, 283
Capital decisions, 61
Cash flows, 64
Cathodes, 220
Change agent, 195
Changing technology, 138
Comparative judgement instrument, 51, 285
Computer-aided design, 134

Computer-aided engineering (CAE), 194
Computer-aided manufacturing (CAM), 10, 56, 134
Computer-aided system, 283
Computer-aided task, 163
Computer crimes, 23
Computer-integrated manufacturing, 10, 150
Computer integrated manufacturing systems (CIMS), 193
Computer performance, 11
Computer revolution, 145
Computers, 3
Computer speed, 56, 165
Contingency leadership, 108
Control System, 153
Corporate culture, 32
Corporate strategic planning, 19
Critical process, 245
Cross impact matrix, 100
Customer surveys, 45

D

Delphi, 100
Design complexity, 58
Design feasibility, 241, 242
Design Interfaces, 251

Design release methodology, 244
Design rules, 23, 125
Design teams, 242
Design verification, 243
Dialogue quality, 58
Direct hours allocation, 71
Domain experts, 124

E

Early manufacturing involvement (EMI), 46, 105, 119
Economy, 41
Educated aspects of technology, 13
Effectiveness, 41
Efficiency, 41
Employee training, 18
Equal productivity curve, 71
Ergonomic issues, 162
Exploratory approach, 99
Expert systems, 123
Expert systems study, 183

F

First technology conference, 13
Fitted curves, 100
Flexible Fabrication System, 149

G

Group technology, 10, 114
Group technology cases, 177
Growth analogy, 100

H

High speed horizontal project, 219
Human operator, 55

I

IBM PC, 146
Ideal innovative manager, 36
Ideal manufacturing goals, 110
Implementation problems, 78
Inalienability, 11
Individual innovations, 36
Individual innovativeness, 35
Indivisibility, 11
Industrial productivity, 2

Inference engine, 124
Information asset security, 23
Information exchange, 34
Information processing, 106
Innovation, 3, 4
 incremental innovations, 4
 radical innovations, 4
 system innovations, 4
Innovation process, 34
Innovation Process Model, 6
Innovation and technology management, 15
Innovative managers, 29
Innovator, 6
Input-output channels, 51
Intangible factors, 188
Internal rate of return, 66
Invention, 3
Islands of technology, 206

J

Japanese-style employment, 202
Job description index, 51
Job satisfaction, 166

K

K-Car, 19
Kinesiology, 45
Knowledge base, 124
Knowledge-based systems, 123
Knowledge engineer, 124

L

Large-scale integration (LSI), 10
Lasers, 10
Linear programming, 73

M

Macroeconomic aspects of technology, 13
Management of technology, 15
Managing technology, 16
Managing technology projects, 76
Manufacturing centers, 148
Manufacturing feasibility, 241
Manufacturing goals, 111
Manufacturing impact assessment, 246
Manufacturing scale-up plan, 249

INDEX

Manufacturing strategy, 219
Manufacturing system integrator, 194
Market positioning, 146
Matrix organization, 107
Microelectronics, 10
Mount structures, 220

N

Next-generation technology, 38
Non-programmed Information, 135
Normative approach, 99

O

Office systems, 143
Operational planning, 97
Optimal design, 59
Organizational structures, 20
Organization change, 39

P

Parallel-phase release, 116
Partial productivity, 25, 67, 69
Participative management, 30
Payback period, 63
Personal computer(s), 17, 143
Phase gate release, 114
Physiological criteria, 51
Planning groups, 32
Policy planning, 97
Preventive maintenance, 93
Process commonality, 249
Process optimization plan, 246
Process qualification plan, 248
Product design, 111
Product development cycles, 252
Product development plan, 244
Product feasibility, 247
Product introduction, 241
Productivity, 67, 165
Product life cycle(s), 18, 110
Product manager, 242, 243
Product release methods, 114
Programmed information, 135
Program plan, 243
Project formation, 80
Project manager, 174
Project phases, 79

Psychological principles, 51
Psychological stress, 56

Q

Quality, 57
Quality improvement, 255
Quality plan(s), 245, 248
Quality of working life, 58, 169
Quantitative techniques, 61

R

Radical innovation, 146
Regression method, 100
Requisition engineering, 151
Research and development, 105–106
Research topics, 257–258
Return on investment, 64
Robotics, 10, 131
Robotics devices, 170

S

Scenarios, 100
Scoring models, 72
S-curves, 18
Semiconductor electronics, 144
Shop-floor language, 212
Single-trend extrapolation, 100
Skeletal system, 47
Skill development, 56
Social aspects of technology, 12
Solutions selling, 211
Strategic planning, 97
Substitution, 100
Supervisory performance, 37
Surface mounted technology, 17
System designers, 23
System improvement, 45
System planning, 45
System productivity, 57
System selection, 45
Systems Integrators (SI), 194
System testing, 45

T

Tactical planning, 97
TASDOC, 43

Task complexity, 58, 165
Task-oriented leader, 108
Technical change, 3
Technical communications, 120
Technical roles, 109
Technological breakthroughs, 34
Technological change, 2
Technological discontinuities, 16, 18
Technological entrepreneur, 30
Technological forecasting, 96, 99
Technological innovation, 1, 255
Technological invention, 4
Technological policy, 32
Technological progress, 12, 85
Technological progress factor, 87
Technology, 10
Technology-Aided Task Impact Model (TATIM), 24
Technology assessment, 93
Technology discontinuities, 187
Technology gaps, 3
Technology implementation, 75
Technology justification, 61, 62
Technology leader, 146
Technology leadership, 143
Technology life cycle, 21
Technology management, 14, 145
Technology management challenges, 20
Technology management mechanisms, 33
Technology policy, 29–32
Technology ratings, 157
Technology report, management of, 34
Technology strategy, 31, 39
Technology transfer, 93
Technology transfer methods, 94
Technology trends, 10

Technology Utilization, 91, 92
TINOP, 135
Total automation, 148
Total factor productivity, 25, 67
Total productivity, 25, 57
Transfer agents, 93
Transistors per chip, 11
Tube, 220
Turnkey operation, 199

U

Unweighted scoring model, 72
User-friendly systems, 56
User strategic plan, 203
Users of technology, 56, 59

V

Value analysis, 112
Value engineering, 91
Variable encoding, 101
Vendor internships, 202
Verbal communication, 109
Very large-scale integration, (VLSI), 10

W

Weighted scoring model, 72
Winds of change, 222
Word processing, 145
Worker's right, 53
Workflow, 210
Working memory, 124
Workmaster console, 149
Workplace design, 165